T0338516

VIRUS INFECTIONS
AND DIABETES MELLITUS

DEVELOPMENTS IN MEDICAL VIROLOGY

Yechiel Becker, Series Editor
Julia Hadar, Managing Editor

Levine, P.H. (ed.) *Epstein-Barr Virus and Associated Diseases* (1985)
Becker, Y. (ed.) *Virus Infections and Diabetes Mellitus* (1987)

DEVELOPMENTS IN MOLECULAR VIROLOGY

Yechiel Becker, Series Editor
Julia Hadar, Managing Editor

Becker, Y. (ed.) *Herpesvirus DNA* (1981)
Becker, Y. (ed.) *Replication of Viral and Cellular Genomes* (1983)
Becker, Y. (ed.) *Antiviral Drugs and Interferon: the Molecular Basis of Their Activity* (1983)
Kohn, A. and Fuchs, P. (eds.) *Mechanisms of Viral Pathogenesis from Gene to Pathogen* (1983)
Becker, Y. (ed.) *Recombinant DNA Research and Viruses. Cloning and Expression of Viral Genes* (1985)
Feitelson, M. (author) *Molecular Components of Hepatitis B Virus* (1985)
Becker, Y. (ed.) *Viral messenger RNA: Transcription, Processing, Splicing and Molecular Structure* (1985)
Doerfler, W. (ed.) *Adenovirus DNA: the Viral Genome and Its Expression* (1986)

DEVELOPMENTS IN VETERINARY VIROLOGY

Yechiel Becker, Series Editor
Julia Hadar, Managing Editor

Payne, L.N. (ed.) *Marek's Disease* (1985)
Burny, A. and Mammerickx, M. *Enzootic Bovine Leukosis and Bovine Leukemia Virus* (1987)
Becker, Y. (ed.) *African Swine Fever* (1987)

VIRUS INFECTIONS AND DIABETES MELLITUS

edited by

Yechiel Becker
The Hebrew University of Jerusalem
Israel

Martinus Nijhoff Publishing
a member of the Kluwer Academic Publishers Group
Boston/Dordrecht/Lancaster

Distributors

for the United States and Canada: Kluwer Academic Publishers, 101 Philip Drive, Assinippi Park, Norwell, MA 02061

for the UK and Ireland: Kluwer Academic Publishers, MTP Press Limited, Falcon House, Queen Square, Lancaster LA1 1RN, UK

for all other countries: Kluwer Academic Publishers Group, Distribution Centre, P.O. Box 322, 3300 AH Dordrecht, The Netherlands

The figure on the cover is mouse pancreatic Islets stained with monoclonal auto antibodies obtained from the hybridoma reprinted with permission of J.-W. Yoon.

Library of Congress Cataloging-in-Publication Data

Virus infections and diabetes mellitus.

(Developments in medical virology ; 2)
Includes bibliographies and index.
1. Diabetes mellitus—Complications and sequelae.
2. Virus diseases. I. Becker, Yechiel. II. Series.
[DNLM: 1. Diabetes Mellitus—complications. 2. Virus Diseases—complications. WK 840 V821]
RC660.V57 1987 616.4'62 87-23994
ISBN 0-89838-970-4

PRINTED IN THE UNITED STATES.

CONTENTS

CONTRIBUTORS

Felix C. Adi
Bethsaida Clinic & Nursing Home
Box 2211
Enugu
Nigeria

Hanoch Bar-On
Department of Internal Medicine B
Hadassah University Hospital
Ein Kerem
91 120 Jerusalem
Israel

Yechiel Becker
Department of Molecular Virology
Faculty of Medicine
The Hebrew University
91 010 Jerusalem
Israel

Graeme I. Bell
Metabolic Research Unit
University of California San Francisco
San Francisco, California 94143
and Chiron Corporation, 4560 Horton Street
Emeryville, California 94608
Present Address:
Howard Hughes Medical Institute
Department of Biochemistry and Molecular Biology
The University of Chicago
920 East 58th Street
Chicago, Illinois 60637
USA

Otto Braun
Institute of Pathology
University of Vienna Medical School
Spitalgasse 4
1090 Vienna
Austria

Peter Buxbaum
Institute of Pathology
University of Vienna Medical School
Spitalgasse 4
1090 Vienna
Austria

E. Cerasi
Department of Endocrinology and Metabolism
The Hebrew University
Hadassah Medical Center
91010 Jerusalem
Israel

Stuart H. Cohen
Division of Infectious Diseases
Department of Internal Medicine
University of California at Davis
School of Medicine
Sacramento, California
USA

Sallie S. Cook
Box 597
Medical College of Virginia
MCV Station
Richmond, Virginia 23298
USA

P.A.W. Edwards
Department of Pathology
University of Cambridge
Cambridge
United Kingdom

Klaus Helmke
Center for Internal Medicine
Justus Liebig University
6300 Giessen
Federal Republic of Germany

Gabriele Horvat
Institute of Pathology
University of Vienna Medical School
Spitalgasse 4
1090 Vienna
Austria

T. Iwasaki
Department of Pathology
Iwate Medical University
Uchimaru 19-1
Morioka 020
Japan

A. Bennett Jenson
Department of Pathology
Georgetown University
Schools of Medicine and Dentistry
Washington D.C. 20007
USA

George W. Jordan
Division of Infectious Diseases
Department of Internal Medicine
University of California at Davis
School of Medicine
Sacramento, California
USA

Nurit Kaiser
Department of Endocrinology and Metabolism
The Hebrew University
Hadassah Medical Center
91010 Jerusalem
Israel

J.H. Karam
Metabolic Research Unit
University of California San Francisco
San Francisco, California 94143
USA

Ingrid Krisch
Second Department of Surgery
University of Vienna Medical School
Spitalgasse 4
1090 Vienna
Austria

Klaus Krisch
Institut fur Pathologische Anatomie
der Universität Wien
Spitalgasse 4
A-1090 Vienna
Austria

T. Kurata
Department of Pathology
National Institute of Health
Kamiosaki 2-10-35
Shinagawaku
Tokyo 141
Japan

Åke Lernmark
Hagedorn Research Laboratory
Niels Steensensvej 6
DK-2820 Gentofte
Denmark

Roger M. Loria
Department of of Microbiology
and Immunology
Box 678
MCV Station
Richmond, Virginia 23298
USA

H. Markholst
Hagedorn Research Laboratory
Niels Steensensvej 6
DK-2820 Gentofte
Denmark

Ram C. Nayak
Joslin Diabetes Center
Harvard Medical School
1 Joslin Place
Boston, MA 02215
U. S. A.

Nikolaus Neuhold
Institute of Pathology
University of Vienna Medical School
Spitalgasse 4
1090 Vienna
Austria

Takashi Onodera
Laboratory of Immunology
National Institute of Animal Health
Jyosuihoncho 1500
Kodaira
Tokyo 187
Japan

Jim T. Posillico
Brigham and Women's Hospital
Harvard Medical School
Boston, MA 02215
U. S. A.

Harvey Rosenberg
Department of Pathology
and Laboratory Medicine
University of Texas Medical School at Houston
Houston, Texas

Elena Ryder
Instituto de Investigaciones Clínicas,
Facultad de Medicina
Universidad del Zulia
Apartado Postal 1151
Maracaibo
Venezuela

Slavia Ryder
Instituto de Investigaciones Clinicas,
Facultad de Medicina
Universidad del Zulia
Apartado Postal 1151
Maracaibo
Venezuela

N. Sanz
Metabolic Research Unit
University of California San Francisco
San Francisco, California 94143
USA

T. Sata
Department of Pathology
National Institute of Health
Kamiosaki 2-10-35
Shinagawaku
Tokyo 141
Japan

Sathyanarayana Srikanta
Joslin Diabetes Center
Harvard Medical School
1 Joslin Place
Boston, MA 02215
U. S. A.

A. Toniolo
Institute of Microbiology
University of Pisa Medical School
13 Via S. Zeno
56100 Pisa
Italy

K. Xiang
Metabolic Research Unit
University of California San Francisco
San Francisco, California 94143
USA

Ji-Won Yoon
Division of Virology
Department of Microbiology and Infectious Diseases
and Laboratory of Viral and Immunopathogenesis of Diabetes
Julia McFarlane Diabetes Research Unit
The University of Calgary
Health Science Centre
330 Hospital Drive NW
Calgary, Alberta
Canada T2N 4N1

PREFACE

This volume in the series *Developments in Medical Virology* deals with viruses involved in diabetes mellitus, a syndrome with a strong genetic background that causes damage to the regulation of insulin synthesis and function. Viruses were found either to cause or to stimulate diabetes mellitus in man and in animal models. The nature of the role of viruses is described by many of the scientists who participated in the original studies. To complete the picture, chapters were included that deal with the insulin gene, the secondary structure of the proinsulin and insulin receptor polypeptides, pancreatic Langerhans islets, and clinical considerations of the disease.

The aim of *Developments in Medical Virology* is to elucidate processes involving viruses as pathogens of cells and organisms, with special attention to human diseases. A number of volumes will be devoted to viruses affecting specific organs (e.g. brain, liver, etc.), while others will elaborate on the clinical experience in the use of antiviral drugs. The series is published in parallel with *Developments in Molecular Virology*, designed to present an analysis of molecular mechanisms implicated in virus infection and replicative processes. In addition, the series *Developments in Veterinary Virology* provides information on viruses causing diseases in animals, with special emphasis on aspects of interest to veterinarians.

I would like to express my appreciation to all the authors for their contributions to this volume, as well as to Dr. J. Hadar for her valued editorial work, and Mrs. E. Herskovics for excellent secretarial help. The continued assistance and encouragement rendered by the Publisher and Vice-President of Martinus Nijhoff Publishing, Mr. Jeffrey K. Smith, are gratefully acknowledged.

Yechiel Becker
Jerusalem

Human insulin gene, insulin and insulin receptor polypeptides and Langerhans islets

1

A POLYMORPHIC LOCUS NEAR THE HUMAN INSULIN GENE ASSOCIATED WITH INSULIN-DEPENDENT DIABETES MELLITUS

K. XIANG*, N. SANZ*, J.H. KARAM* AND G.I. BELL*†

*Metabolic Research Unit, University of California San Francisco, San Francisco, California 94143 and †Chiron Corporation, 4560 Horton Street, Emeryville, California 94608

ABSTRACT

There is a polymorphic locus near the beginning of the human insulin gene on chromosome 11 whose size is extremely variable. This locus, the insulin gene hypervariable region (HVR), is composed of tandem repeats of a family of related oligonucleotides. Its size varies because of variation in the number of repeats. Population studies indicate that the sizes of the HVR fall into three heterodisperse classes, designated 1, 2 and 3 in order of increasing size and whose frequency varies between racial groups. The common Class 1 allele and the homozygous Class 1 genotype are significantly more frequent ($p < 0.001$) in Caucasians with insulin-dependent diabetes mellitus (IDDM) than in patients with non-insulin-dependent diabetes or non-diabetic controls. The Class 1 HVR allele may be a marker for an allele of a gene which predisposes to IDDM. Although the identity of this diabetogenic locus has yet to be determined, possible candidates include: 1) a gene encoding a beta-cell specific autoantigen; 2) one which determines the susceptibility of the beta cell to viral infection or its response to such infection; and 3) a gene which influences beta cell regeneration.

INTRODUCTION

Diabetes mellitus comprises a heterogenous group of disorders whose etiology seems to be multifactorial with both genetic and environmental factors contributing to its development (1-3).

Type I or insulin-dependent diabetes mellitus (IDDM) is characterized by severe deficiency of insulin secretion due to profound beta cell destruction. These patients require therapy with exogenous insulin to

Becker, Y (ed), Virus Infections and Diabetes Melitus. ISBN 0-89838-970-4.

avoid severe hyperglycemia and ketosis. Although this disorder occurs at any age, its onset is most common in the young and circulating islet cell antibodies are detected in as many as 80-85% of these patients at the onset of their disease. In most cases of IDDM, a positive association with the Class II major histocompatibility complex (MHC) antigens, HLA-DR3 and -DR4, suggests that genetic factors on the short arm of human chromosome 6 (which carries the MHC) contribute to the development of manifest diabetes. However, monozygotic twin studies by Pyke and his colleagues (4) show that genetic factors alone are not sufficient to produce diabetes since as many as 44% of twins in a large series (147 pairs) were discordant for diabetes. Moreover, among Caucasians approximately 50% of non-diabetics are HLA-DR3 or -DR4 and yet only about 0.1% of the population develops IDDM. These observations support the concept that additional genetic determinants and/or environmental factors such as viruses, drugs or toxic chemicals may be required to precipitate the development of IDDM in individuals having HLA-DR3 and/or HLA-DR4 antigens. This type of diabetes is found primarily in Caucasians of northern Europe or their descendents. In the United States, which has a mixed racial population, about 5,500,000 patients are projected to have diabetes mellitus and the prevalence of IDDM in this group is about 10% (5). By contrast, in Scandinavia 20% of the diabetic population have IDDM. In Asia and Africa this form occurs in less than 1% of the diabetic patients.

Type II or non-insulin-dependent diabetes mellitus (NIDDM) is less well-characterized and probably represents an even more heterogenous group of patients than Type I. These are patients with a less severe form of diabetes who do not require insulin therapy for survival. They show no association with HLA antigens and have no demonstrable circulating islet cell antibodies. This type of diabetes develops more frequently in older individuals, of all racial groups, manifesting itself most commonly after the age of 40 years and particularly when obesity is developing or progressing in parallel with advancing age. Concordance for NIDDM in identical twins approaches 100%, indicating that there is a significant genetic factor in the etiology of this form of diabetes (4). The clinical and biochemical features of NIDDM suggest that components of the insulin-producing beta cell as well as of the insulin responsive target cells could be defective. For example, the synthesis of a mutant insulin or proinsulin molecule or abnormalities in the regulation of insulin gene expression that result in reduced insulin production could conceivably contribute to the disease (6). The abnormal pattern of insulin secretion seen in these patients could be a consequence of lower insulin content of the secretory granules of their beta cells, reduced beta cell mass or altered responsiveness of the beta cells to glucose and other secretory stimuli. The impaired insulin action (7,8) has been attributed to altered insulin binding to its receptors on target tissues (liver, muscle and fat cells) as well as to a pronounced defect in the intracellular action of insulin at a post-receptor level in these tissues.

THE INSULIN GENE REGION OF CHROMOSOME 11

The isolation and characterization of the human insulin gene has allowed its role in the etiology of diabetes to be critically examined (9-12). The human insulin (INS) gene is located on the short arm of chromosome 11 in band p15 (13). The insulin-like growth factor II (IGF2) gene and the genes of the β-globin complex (HBB) as well as the proto-oncogene, c-Harvey-ras 1 (HRAS1) are located in the same band, a region of ~ 20×10^6 base pairs (bp). The IGF2 gene is immediately adjacent to INS; the genes have the same orientation and are separated by less than 12.5×10^3 bp (Fig. 1). Linkage studies indicate that the HBB and HRAS1 genes are on opposite sides of INS/IGF2; HBB is located ~ 10 centiMorgans (cM) (~10×10^6 bp) closer to the centromere than INS/IGF2 and HRAS1 is ~ 3 cM (3×10^6 bp) nearer the telomere.

The insulin genes of diabetic and non-diabetic subjects were examined using the Southern blotting technique for restriction fragment length polymorphisms (RFLPs) which might be unique to diabetic individuals and thus indicate a role for this gene in the etiology of diabetes (14,15). These studies revealed a region of variable size, the insulin gene hypervariable region (HVR) (Fig. 1), located 365 bp before the start of insulin mRNA synthesis (Fig. 2). Population studies indicated that based upon their size (Fig. 3) alleles of the HVR can be divided into three main classes, designated Class 1, 2 and 3, having an average size of 570, 1320 and 2470 bp. The sequence of the HVR (Fig. 2) indicates that it is composed of multiple tandem repeats of 14-15 bp segments

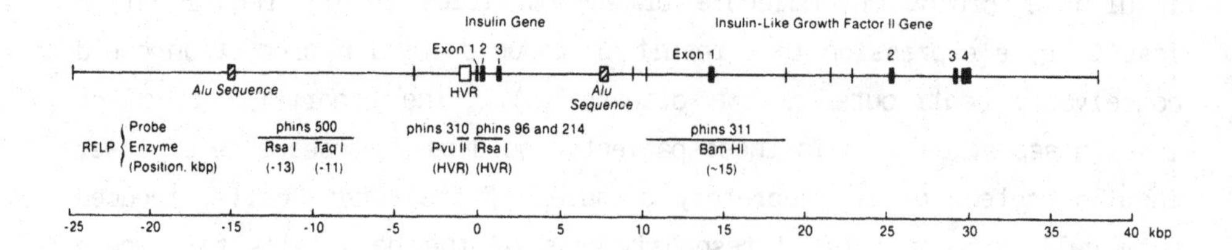

Fig. 1. Map of the INS/IGF2 region. The positions of exons, the hypervariable region (HVR) and the dispersed middle repetitive Alu sequences are indicated. The vertical lines in the upper part of the figure denote EcoRI sites. Coordinates are presented relative to the start of transcription of the insulin gene. The locations of various RFLPs are noted.

whose consensus sequence is ACAGGGGTGTGGGG. Thus, Class 1, 2 and 3 alleles have an average of 40, 95 and 170 tandem repeats. There can also be small differences in the size of alleles of the same class due to the presence of alleles which contain slightly different numbers of repeats (11). In addition, alleles of the same class can also differ in nucleotide sequence and thus there may be an infinite number of HVRs of different sizes and sequences. The HVR is inherited in a Mendelian fashion and has the same organization in DNA prepared from an insulin-producing pancreatic islet cell adenoma and white-blood cells of the same individual.

RACIAL DIFFERENCES IN THE FREQUENCIES OF HVR ALLELES

There are differences in the frequency of Class 1, 2 and 3 alleles among racial groups (Fig. 4). Class 1 alleles are found in all racial groups; their frequency varies from 0.57 in African Blacks to 0.98 in Orientals (Chinese and Japanese). Class 2 alleles occur almost exclusively in Blacks; their frequency in Caucasians is less than 0.01 and they have not been reported in any of the other racial groups. The frequency of Class 3 alleles is lowest in Orientals (0.02) and highest in Caucasians (0.34), as a consequence of the differences in the frequency of the three classes of alleles of the HVR, Blacks possess the greatest genotypic diversity and Orientals the least. The differences in allelic

```
                                                                          Hypervariable Region
CCCCTCCCTCACTCCCACTCTCCCACCCCCACCACCTTGGCCCATCCATGGCGGCATCTTGGGCCATCCGGGACTGGGG (ACAGGGGTCCTGGGGACAGGGGTCCGGGGACAGGGTCCTGGG  -806

GACAGGGGTGTGGGGACAGGGGTCTGGGGACAGGGGTGTGGGGACAGGGGTGTGGGGACAGGGGTCTGGGGACAGGGGTGTGGGGACAGGGGTCCGGGGACAGGGGTGTGGGGACAGGGGT  -685

CTGGGGACAGGGGTGTGGGGACAGGGGTGTGGGGACAGGGGTCTGGGGACAGGGGTGTGGGGACAGGGGTCCTGGGGACAGGGGTGTGGGGACAGGGGTGTGGGGACAGGGGTGTGGGGAC  -564

AGGGGTGTGGGGACAGGGGTCCTGGGGATAGGGGTGTGGGGACAGGGGTGTGGGGACAGGGGTCCCGGGGACAGGGGTGTGGGGACAGGGGTGTGGGGACAGGGGTCCTGGGGACAGGGGT  -443

CTGAGGACAGGGGTGTGGGCACAGGGGTCCTGGGGACAGGGGTCCTGGGGACAGGGGTCCTGGGGACAGGGGTCTGGGG) ACAGCAGCGCAAAGAGCCCCGCCCTGCAGCCTCCAGCTCTCC  -322

TGGTCTAATGTGGAAAGTGGCCCAGGTGAGGGCTTTGCTCTCCTGGAGACATTTGCCCCCAGCTGTGAGCAGGGACAGGTCTGGCCACCGGGCCCCTGGTTAAGACTCTAATGACCCGCTG  -201

GTCCTGAGGAAGAGGTGCTGACGACCAAGGAGATCTTCCCACAGACCCAGCACCAGGGAAATGGTCCGGAAATTGCAGCCTCAGCCCCCAGCCATCTGCCGACCCCCCCACCCCAGGCCCT   -80

                                                                          Exon 1
AATGGGCCAGGCGGCAGGGGTTGACAGGTAGGGGAGATGGGCTCTGAGACTATAAAGCCAGCGGGGGCCCAGCAGCCCTC AGCCCTCCAGGACAGGCTGCATCAGAAGAGGCCATCAAGCA    41

 Intron A
G GTCTGTTCCAAGGGCCTTTGCGTCAGGTGGGCTCAGGGTTCCAGGGTGGCTGGACCCCAGGCCCCAGCTCTGCAGCAGGGAGGACGTGGCTGGGCTCGTGAAGCATGTGGGGGTGAGCCC  162

                                                                         -24             -20
                                                   Exon 2                Met Ala Leu Trp Met Arg Leu Leu Pro Leu Leu
AGGGGCCCCAAGGCAGGGCACCTGGCCTTCAGCCTGCCTCAGCCCTGCCTGTCTCCCAG ATCACTGTCCTTCTGCC ATG GCC CTG TGG ATG CGC CTC CTG CCC CTG CTG    271

    -10                                 1                                       10
Ala Leu Leu Ala Leu Trp Gly Pro Asp Pro Ala Ala Ala Phe Val Asn Gln His Leu Cys Gly Ser His Leu Val Glu Ala Leu Tyr Leu
GCG CTG CTG GCC CTC TGG GGA CCT GAC CCA GCC GCA GCC TTT GTG AAC CAA CAC CTG TGC GGC TCA CAC CTG GTG GAA GCT CTC TAC CTA    361

    20                                  30
Val Cys Gly Glu Arg Gly Phe Phe Tyr Thr Pro Lys Thr Arg Arg Glu Ala Glu Asp Leu Gln V  Intron B
GTG TGC GGG GAA CGA GGC TTC TTC TAC ACA CCC AAG ACC CGC CGG GAG GCA GAG GAC CTG CAG G  GTGAGCCAACCGCCCATTGCTGCCCCTGGCCGCCCC  461

CAGCCACCCCCTGCTCCTGGCGCTCCCACCCAGCATGGGCAGAAGGGGGCAGGAGGCTGCCACCCAGCAGGGGGTCAGGTGCACTTTTTTAAAAAGAAGTTCTCTTGGTCACGTCCTAAAA  582

GTGACCAGCTCCCTGTGGCCCAGTCAGAATCTCAGCCTGAGGACGGTGTTGGCTTCGGCAGCCCCGAGATACATCAGAGGGTGGGCACGCTCCTCCCTCCACTCGCCCCTCAAACAAATGC  703

CCCGCAGCCCATTTCTCCACCCTCATTTGATGACCGCAGATTCAAGTGTTTTGTTAAGTAAAGTCCTGGGTGACCTGGGGTCACAGGGTGCCCCACGCTGCCTGCCTCTGGGCGAACACCC  824

CATCACGCCCGGAGGAGGGCGTGGCTGCCTGCCTGAGTGGGCCAGACCCCTGTCGCCAGCCTCACGGCAGCTCCATAGTCAGGAGATGGGGAAGATGCTGGGGACAGGCCCTGGGGAGAAG  945

TACTGGGATCACCTGTTCAGGCTCCCACTGTGACGCTGCCCCGGGGCGGGGGAAGGAGGTGGGACATGTGGGCGTTGGGGCCTGTAGGTCCACACCCAGTGTGGGTGACCCTCCCTCTAAC  1,066

CTGGGTCCAGCCCGGCTGGAGATGGGTGGGAGTGCGACCTAGGGCTGGCGGGCAGGCGGGCACTGTGTCTCCCTGACTGTGTCCTCCTGTGTCCCTCTGCCTCGCCGCTGTTCCGGAACCT  1,187

                        Exon 3                                        50                                      60
                        al Gly Gln Val Glu Leu Gly Gly Gly Pro Gly Ala Gly Ser Leu Gln Pro Leu Ala Leu Glu Gly Ser Leu
GCTCTGCGCGGCACGTCCTGGCAG TG GGG CAG GTG GAG CTG GGC GGG GGC CCT GGT GCA GGC AGC CTG CAG CCC TTG GCC CTG GAG GGG TCC CTG  1,282

                    70                                  80          86
Gln Lys Arg Gly Ile Val Glu Gln Cys Cys Thr Ser Ile Cys Ser Leu Tyr Gln Leu Glu Asn Tyr Cys Asn AM
CAG AAG CGT GGC ATT GTG GAA CAA TGC TGT ACC AGC ATC TGC TCC CTC TAC CAG CTG GAG AAC TAC TGC AAC TAG ACGCAGCCTGCAGGCAGCCCC  1,378

                          Polyadenylation Site
ACACCCGCCGCCTCCTGCACCGAGAGAGATGGAATAAAGCCCTTGAACCAGC CCTGCTGTGCCGTCTGTGTGTCTTGGGGGCCCTGGGCCAAGCCCCACTTCCCGGCACTGTTGTGAGCCC  1,499
```

Fig. 2. Sequence of the human insulin gene and flanking Class 1 HVR allele having 34 tandem repeats. The start of insulin gene transcription (i.e., the beginning of exon 1) has been designated as nucleotide 1. The number of the nucleotide at the end of each line is noted.

frequencies observed between African and American Blacks presumably reflects the presence of Caucasian genes in the latter group. The similarity in frequency between Caucasians from northern Europe and North America (this group has a diverse ethnic background and includes individuals from all parts of Europe and the Middle East), suggests that there is probably little, if any, ethnic variation. The differences in alle-

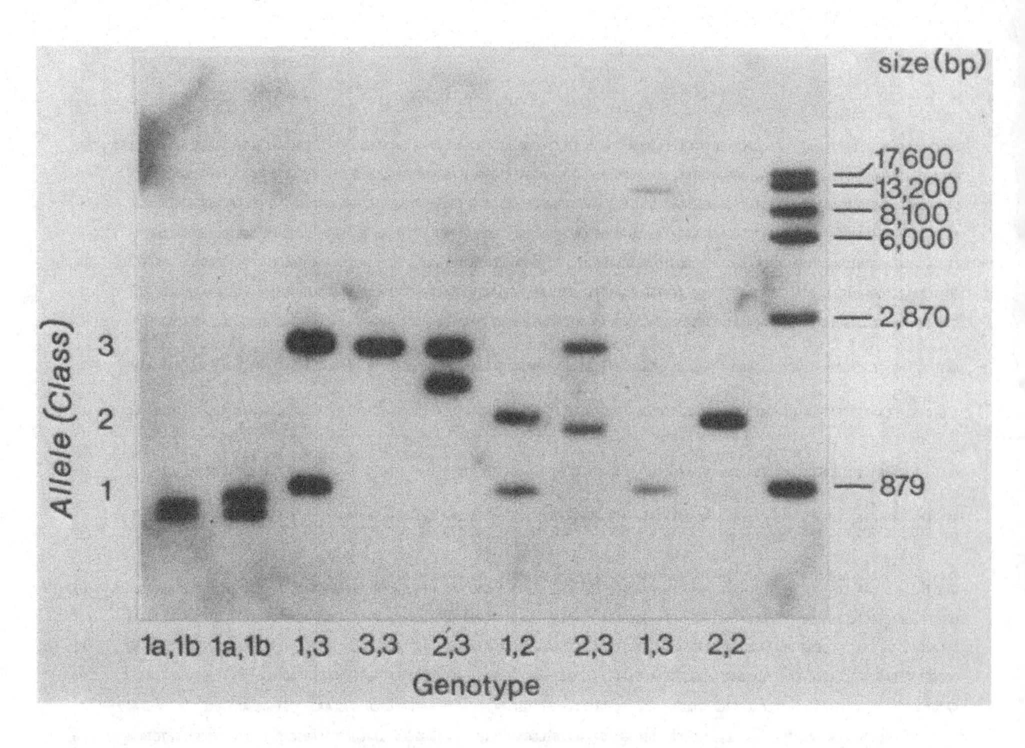

Fig. 3. Southern blot showing variation in size of the HVR. The patterns observed after digesting lymphocyte DNA of nine individuals with PvuII and probing with phins310 are shown. The genotype of the subject is indicated at the bottom of the lane; a and b indicate that this individual has two alleles of the same class which can be distinguished by their size. The right lane contains size markers.

lic frequencies between racial groups indicate that in studies of the HVR, it is important that affected subjects and controls are the same race.

THE HYPERVARIABLE REGION AND DIABETES

The proximity of the HVR to the insulin gene and its unusual sequence characteristics suggested that it might affect insulin gene expression and thereby contribute to the deficient insulin secretory reserve observed in NIDDM patients. The frequencies of the three HVR classes in diabetic and non-diabetic patients were compared to determine if it might be a molecular marker for this disease.

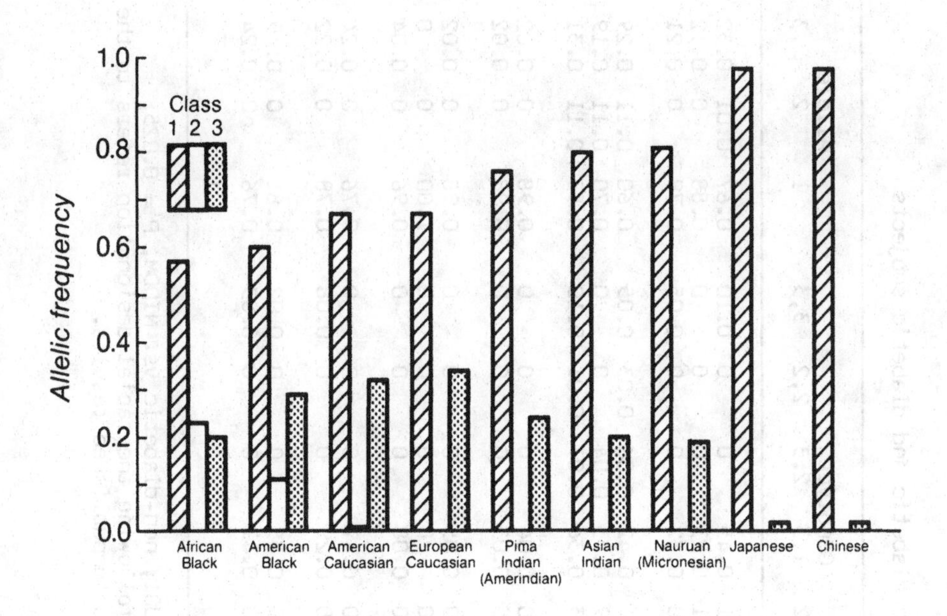

Fig. 4. The frequency of Class 1, 2 and 3 alleles of the HVR in non-diabetic individuals of different racial and ethnic groups. This summary was prepared from the data presented in references 14-21 and our unpublished data for Chinese.

These studies indicated that there is no allelic class that is present in only diabetic or non-diabetic subjects (Table 1) and thus the HVR cannot be used to identify diabetic or pre-diabetic subjects (14,22). Associations of the HVR and diabetes have been studied in Caucasians in North America [San Francisco (14,22) and St. Louis (23)] and Europe [Copenhagen (24) and London (19)], American Blacks (18), Orientals [Chinese (14) and Japanese (14,20,21)], Pima Indians (16) and inhabitants of the Micronesian Island of Nauru (15). The latter two groups are particularly interesting because of the very high frequency of NIDDM among adults (30%). A comparison of the HVR genotypic and allelic-frequencies between non-diabetic and NIDDM groups indicate that there are no significant differences in frequency within any of the racial or ethnic groups (the relevance of such comparisons in Orientals is questionable given the very low frequency of non-Class 1 alleles in this group).

Table 1. HVR genotypic and allelic frequencies in non-diabetic and diabetic subjects

Race	Group	Genotype						Allele		
	(No. examined)	1,1	1,2	1,3	2,3	2,2	3,3	1	2	3
Caucasian (14)	Non-diabetics (83)	0.45	0.01	0.45	0	0	0.10	0.67	0.01	0.32
	IDDM (113)	0.76	0.01	0.23	0	0	0	0.88	0	0.12
	NIDDM (76)	0.63	0	0.32	0	0	0.05	0.79	0	0.21
American Blacks (18)	Non-diabetics (132)	0.33	0.11	0.42	0.06	0.03	0.05	0.60	0.11	0.29
	IDDM (27)	0.44	0.19	0.33	0.04	0	0	0.70	0.11	0.19
	NIDDM (154)	0.34	0.13	0.36	0.07	0.01	0.10	0.58	0.11	0.31
Chinese	Non-diabetics (46)	0.96	0	0.04	0	0	0	0.98	0	0.02
	NIDDM (33)	0.97	0	0.03	0	0	0	0.98	0	0.02
Japanese (20)	Non-diabetics (64)	0.95	0	0.05	0	0	0	0.98	0	0.02
	IDDM (39)	1.00	0	0	0	0	0	1.00	0	0
	NIDDM (51)	0.92	0	0.08	0	0	0	0.96	0	0.04
Pima Indians (16)	Non-diabetics (38)	0.53	0	0.47	0	0	0	0.76	0	0.24
	NIDDM (49)	0.63	0	0.29	0	0	0.08	0.78	0	0.22
Micronesians (15)	Non-diabetics (60)	0.63	0	0.35	0	0	0.02	0.81	0	0.19
	NIDDM (58)	0.55	0	0.41	0	0	0.03	0.76	0	0.24

Statistics: Caucasian; non-diabetic vs. IDDM, $P < 0.001$; non-diabetic vs. NIDDM, $P = 0.025$; IDDM vs. NIDDM, $P < 0.05$. The number in parentheses following the racial designation refers to the published source of the data. The Chinese data are our unpublished results.

Initial reports had suggested that there was an association of the larger Class 3 alleles with NIDDM (24,25,26), however recent studies do not support this conclusion. The association reported by Rotwein and his colleagues (25,26) was based upon a comparison of allelic frequencies between racially mixed groups and is not confirmed when diabetic and NIDDM subjects of the same racial group are compared (23). Although Danish investigators had reported an association of two Class 3 alleles with NIDDM in an earlier publication, they subsequently concluded that the Class 3 allele is a marker for the development of atherosclerosis and they attribute their previously observed association with NIDDM to the increased prevalence of atherosclerosis in their NIDDM subjects (27). In contrast to the above groups, in our Caucasian NIDDM patients, we observed a significantly higher frequency of Class 1 alleles and homozygous Class 1 genotypes ($P=0.025$) compared to non-diabetic controls (14). This has not been confirmed and could reflect the difficulty of accurately classifying diabetic patients; for example, subjects with a mild form of IDDM, having only partial beta-cell destruction, may have clinical features warranting classification as NIDDM. This has recently been documented (28) wherein as many as 14% of patients classified as NIDDM were found to have islet cell antibodies suggesting a diagnosis of mild IDDM.

While the aggregate data in Caucasians indicate that the HVR does not directly predispose an individual to NIDDM, the heterogeneity of this disease and the observed differences in the frequency of HVR alleles between different NIDDM groups suggests that there might be associations with subgroups of Caucasian NIDDM subjects.

The frequencies of the three classes of HVR alleles were also compared between IDDM subjects and NIDDM or non-diabetic control groups. In Caucasians with IDDM, the frequencies of HVR Class 1 alleles and the homozygous Class 1 genotype were significantly higher ($P<0.001$) than in non-diabetic or NIDDM groups. This observation has been confirmed (19) and is also supported by analysis of the combined data obtained from the various studies (14). The frequency of the Class 1 allele also tends to be higher in American Blacks with IDDM than in non-diabetic or NIDDM subjects (18), although the difference is not significant; this could reflect admixture of Black and Caucasian genes in this group.

The HVR has also been studied as a marker for other diseases. However, the reported associations are controversial [an increased frequency of Class 3 alleles in patients with atherosclerosis (27,29)] or unconfirmed [the association of Class 3 alleles with elevated glycosylated hemoglobin levels (30) or diabetic hypertriglyceridemia (31); Class 1 alleles with impaired glucose tolerance in acromegalic subjects (32)]. HVR alleles have also been studied as a possible marker for islet beta-cell function, but no association with insulin secretory response was observed (33). Finally the HVR has been used as marker to study the inheritance of maturity-onset diabetes of the young (MODY), a rare form of diabetes which does not require insulin therapy and is characterized by an autosomal dominant pattern of inheritance and a low frequency of diabetic complications. The insulin gene did not segregate with this disorder in any of the four families that were examined indicating that the gene defect in these individuals is not close to the insulin gene (34-36).

CONCLUSION

The association of Class 1 alleles of the insulin gene HVR and IDDM was unexpected as the absolute insulin deficiency in these patients is due to the destruction of the insulin-producing beta cells rather than an abnormality in insulin gene expression. The beta-cell loss in IDDM is believed to be the result of a specific autoimmune response directed against the beta cell (37-40). The association of HLA-DR3 and -DR4 antigens with IDDM suggests that they, or the products of genes linked to those for HLA-DR3 and -DR4, influence the abnormal immunological reaction against the beta cell. These antigens are present in more than 90% of IDDM patients. They are also prevalent in ~50% of non-diabetic subjects, yet only about 0.1% of the population develops IDDM. In fact, it seems to be a general rule with HLA and disease associations that only a small proportion of individuals who have the particular HLA antigen actually develop the disease with which it is associated (41). The reasons for this are not clearly understood but there are four possible explanations: 1) the disease is heterogeneous and the association only occurs in a subset of individuals; 2) the association is with a diabetes susceptibility gene at a closely linked locus whose frequency is lower than that of the HLA locus to which it is linked and as a consequence only a fraction of the chromosomes carrying the HLA allele will also carry the susceptibility gene; 3) other non-HLA genes also contribute to diabetes susceptibility; and 4) environmental factors may determine which individuals develop the overt disease. Clearly, the pathophysiology and heterogeneity of IDDM suggests that each of these explanations may be relevant to its etiology. They could also explain why only a fraction of the individuals who have a Class 1 HVR become diabetic.

The role of the HVR in the etiology of IDDM is speculative. It is unlikely that it is the susceptibility locus as it probably does not encode a protein (11). The proximity of the HVR to the insulin gene suggests that it might confer some control over the quantity of insulin produced. However, available data provides no evidence that HVR alleles affect insulin synthesis or secretion. Nor is it obvious, how differences in insulin production could contribute to the immunological destruction of the beta cell. Thus, it seems likely that the Class 1 HVR is a marker for an allele of a closely-linked gene which predisposes to IDDM. Candidate susceptibility genes include those encoding beta-cell specific antigens (i.e., the target antigens) to which the autoimmune response is directed as well as loci which might affect the sensitivity or response of the beta cell to the environmental agent. For example, the Class 1 HVR could be linked to a locus encoding a membrane protein, a variant of which shares an antigenic determinant with a common virus. In this case, the autoimmune destruction of the beta cell is a consequence of the cross-reactivity of a surface protein on normal beta cells with antibodies and cytotoxic cells directed against an infecting virus. The diabetogenic locus near the INS gene might determine the susceptibility of the beta cell to viral infection. In fact, a non-MHC locus has been identified in mice which determines susceptibility to virally-induced diabetes (42), possibly by controlling the number of molecules on the beta cell surface that can function as viral receptors; sensitive strains have more receptors than those that are resistant. The HVR could also be a marker for a locus which determines the response of the beta cell to viral infection, for example by affecting ability of antiviral agents like interferon, to inhibit replication (43), or for a gene which could possibly influence beta-cell regeneration, e.g., IGF-II.

The notion that this susceptibility gene somehow determines the extent of viral damage or beta-cell mass is particularly attractive. It suggests that there may be a spectrum of beta cell destruction: in some individuals there would be minimal loss of cells and no impairment of function; in others there could be a slight to moderate reduction in beta-cell mass which could be a predisposing factor for developing NIDDM; finally, the most susceptible individuals, those who also express HLA-DR3 and -DR4 antigens, would experience severe beta-cell loss and have overt IDDM. This model predicts that there might be a weak association between Class 1 alleles and NIDDM which is what we have observed.

Genetic markers linked to susceptibility genes for IDDM have now been identified on human chromosomes 6 and 11. In addition, the clinical features of the disease suggest that other loci which may be relevant to its pathophysiology have yet to be identified; for example, there may be genes which determine the severity of the various complications experienced by individuals with IDDM. Our task is now to specifically identify these susceptibility genes and determine how they interact and contribute to the etiology of this disease.

ACKNOWLEDGEMENTS

This work was supported by the Dorothy Frank Research Fund and a gift from Mr. and Mrs. Jay Frankel. Dr. K. Xiang received support from the World Health Organization, program number G4 107.

REFERENCES

1. Rotter, J.I. and Rimoin, D.L. Am. J. Med. 70: 116-126, 1981.
2. Leslie, R.D.G. and Pyke, D.A. In: The Diabetes Annual 1 (Eds. K.G.M.M. Alberti and L.P. Krall), Elsevier, Amsterdam, 1985, pp. 53-66.
3. Rudiger, H.W. and Dreyer, M. Hum. Gene. 63:100-110, 1983.
4. Barnett, A.H., Eff, C., Leslie, R.D.G. and Pyke, D.A. Diabetologia 20:87-93, 1981.
5. Herman, W.H., Sinnock, P., Brenner, E., Brimberry, J.L., Langford, D., Nakashima, A., Sepe, S.J., Teutsch, S.M. and Mazze, R.S. Diabetes Care 7:367-371, 1984.
6. Tager, H.S. Diabetes 33:693-699, 1984.
7. Gerich, J.E. J. Lab. Clin. Med. 103:497-505, 1984.
8. Kolterman, O., Gray, R., Griffin, J., Burstein, P., Insel, J., Scarlett, J. and Olefsky, J. J. Clin. Invest. 68:957-969, 1981.

9. Bell, G.I., Pictet, R.L., Rutter, W.J., Cordell, B., Tischer, E. and Goodman, H.M. Nature 284:26-32, 1980.
10. Bell, G.I., Pictet, R. and Rutter, W.J. Nucleic Acids Res. 8:4091-4109, 1980.
11. Bell, G.I., Selby, M.J. and Rutter, W.J. Nature 295:31-35, 1982.
12. Bell, G.I., Gerhard, D.S., Fong, N.M., Sanchez-Pescador, R. and Rall, L.B. Proc. Natl. Acad. Sci. USA 82:6450-6454, 1985.
13. Grzeschik, K.-H. and Kazazian, H.H. Cytogenet. Cell Genet. 40:179-205, 1985.
14. Bell, G.I., Horita, S. and Karam, J.H. Diabetes 33:176-183, 1984.
15. Serjeantson, S.W., Owerbach, D., Zimmet, P., Nerup, J. and Thoma K. Diabetologia 25:13-17, 1983.
16. Knowler, W.C., Pettitt, D.J., Vasquez, B., Rotwein, P.S., Andreone, T.L. and Permutt, M.A. J. Clin. Invest. 74:2129-2135, 1984.
17. Williams, L.G., Jowett, N.I., Vella, M.A., Humphries, S. and Galton, D.J. Hum. Genet. 71:227-230, 1985.
18. Elbein, S., Rotwein, P., Permutt, M.A., Bell, G.I., Sanz, N. and Karam, J.H. Diabetes 34:433-439, 1985.
19. Hitman, G.A., Tarn, A.C., Winter, R.M., Drummond, V., Williams, L.G., Jowett, N.I., Bottazzo, G.F. and Galton, D.J. Diabetologia 28:218-222, 1985.
20. Awata, T., Shibasaki, Y., Hirai, H., Okabe, T., Kanazawa, Y. and Takaku, F. Diabetologia 28:911-913, 1985.
21. Haneda, M., Kobayashi, M., Maegawa, H. and Shigeta, Y. Diabetes 35:115-118, 1986.
22. Bell, G.I., Karam, J.H. and Rutter, W.J. Proc. Natl. Acad. Sci. USA 78:5759-5763, 1981.
23. Permutt, M.A., Andreone, T., Chirgwin, J. Elbein, S., Rotwein, P. and Orland, M. In: Comparison of Type I and Type II Diabetes (Eds. M. Vranic, C.H. Hollenberg and G.Steiner), Plenum Press, New York, 1985, pp.89-106.
24. Owerbach, D. and Nerup, J. Diabetes 31:275-277, 1982.
25. Rotwein, P., Chyn, R., Chirgwin, J., Cordell, B., Goodman, H.M. and Permutt, M.A. Science 213:1117-1120, 1981.
26. Rotwein, P.S., Chirgwin, J., Province, M., Knowler, W.C., Pettitt, D.J., Cordell, B., Goodman, H.M. and Permutt, M.A. N. Engl. J. Med. 308:65-71, 1983.
27. Mandrup-Poulsen, T., Mortensen, S.A., Meinertz, H., Owerbach, D., Johansen, K., Sorensen, H. and Nerup, J. Lancet i:250-252, 1984.
28. Groop, L.C., Bottazzo, G.F. and Doniach, D. Diabetes 35:237-241, 1986.
29. Rees, A., Stocks, J., Williams, L.G., Caplin, J.L., Jowett, N.I., Camm, A.J. and Galton, D.J. Atherosclerosis 58:269-275, 1985.
30. Owerbach, D., Billesbolle, P., Poulsen, S. and Nerup, J. Lancet i:880-883, 1982.
31. Jowett, N.I., Williams, L.G., Hitman, G.A. and Galton, D.J. Br. Med. J. 288:96-99, 1984.

32. Hitman, G.A., Katz, J., Lytras, N., Jowett, N.I., Wass, J.A.H., Besser, G.M. and Galton, D.J. Clin. Endocrin. 23:817-822, 1985.
33. Permutt, M.A., Rotwein, P., Andreone, T., Ward, W.K. and Porte, D. Diabetes, 34:311-314, 1985.
34. Bell, J.I., Wainscoat, J.S., Old, J.M., Chlouverakis, C., Keen, H., Turner, R.C. and Weatherall, D.J. Br. Med. J. 286:590-592, 1983.
35. Owerbach, D., Thomsen, B., Johansen, K., Lamm, L.U. and Nerup, J. Diabetologia 25:18-20, 1983.
36. Andreone, T., Fajans, S., Rotwein, P., Skolnick, M. and Permutt, M.A. Diabetes 34:108-114, 1985.
37. Cahill, G.F. and McDevitt, H.O. N. Engl. J. Med. 304:1454-1465, 1981.
38. Lernmark, A. Diabaetologia 28:195-203, 1985.
39. Eisenbarth, G.S., N. Engl. J. Med. 314:1360-1368, 1986.
40. Bottazzo, G.F. Diabetic Medicine 3:119-130, 1986.
41. Bodmer, W.F. and Bodmer, J.G., Br. Med. Bull. 34:309-316, 1978.
42. Onodera, T., Yoon, J.-W., Brown, K.S. and Notkins, A.L. Nature 274:693-696, 1978.
43. Craighead, J.E. Am. J. Med. 70:127-134, 1981.

2

COMPUTERIZED CHOU-FASMAN SECONDARY STRUCTURE ANALYSES OF PROINSULIN, INSULIN AND INSULIN RECEPTOR POLYPEPTIDES

YECHIEL BECKER

Department of Molecular Virology, Faculty of Medicine, The Hebrew University, 91 010 Jerusalem, Israel

ABSTRACT

The secondary structures of proinsulin, insulin and the precursor molecules of the insulin receptor (INSR) were determined by a computer program that utilizes the Chou-Fasman and Robson-Garnier calculations. These polypeptides have a signal peptide at the N terminus that is cleaved during transfer through the cell membrane. The insulin receptor polypeptide contains a domain of 17 hydrophobic amino acids in the β subunit which is inserted into the cell membrane. Two thirds of the polypeptide, the entire α subunit, and part of the β subunit are on the outside of the cell, and the rest is located in the cytoplasm, having a reported function of tyrosine kinase. The β subunit of the INSR is homologous in primary and secondary structure to that of the oncogenes of the tyrosine kinase family. The computer program GAP allowed comparison of the primary amino acid sequence of insulin with that of the insulin-like growth factors I and II revealing partial homology. A domain in each of the α and β subunits of the INSR was found to be partially homologous with the insulin B chain and proinsulin.

Becker, Y (ed), Virus Infections and Diabetes Melitus. ISBN 0-89838-970-4.

INTRODUCTION

The human gene that codes for insulin (1), was found to be located in the short arm of chromosome 11 in band p15 (2). The proinsulin gene consists of three exons and two introns coding for a proinsulin polypeptide110 amino acids (aa) in length. A repeated nucleotide sequence region of variable size (named the insulin hypervariable region (HVR)) is located 365 base pairs (bp) before the start of proinsulin mRNA (1). The insulin-like growth factor II (IGF2) gene is located downstream from the insulin gene. The insulin gene functions in the beta cells in the pancreas where mRNA for the proinsulin peptides is produced. In these cells, cytoplasmic granules contain the proinsulin molecules which, due to their ability to fold into secondary and tertiary conformations contain segments of the polypeptide amino acids [(aa) 25-54 in the B chain and 90-110 in the A chain] that are cross-linked by the formation of S-S bonds between two cysteine molecules. Proteolytic cleavage releases a peptide segment containing aa 57-87 of the proinsulin (C peptide). During transfer of the proinsulin chain through the membrane into the cytoplasmid granule, the signal peptide consisting of aa 1-24 is cleaved off, and the active insulin molecule is released from the beta cells of the pancreas.

The rapid metabolic effects of insulin are initiated by the interaction of the latter with specific insulin receptor (INSR) molecules present on the surface of cells in all tissues and organs in the body (3). The insulin receptor to which the insulin molecule binds has, upon purification, a molecular mass (M_r) of 350,000 to 400,000, and is made up of two α and two β subunit polypeptides. The isolation and cloning of the human gene for the insulin receptor (3) provided the complete sequence of the insulin receptor polypeptide, as well as clues for specific domains in the polypeptide responsible for passage of the 1370 aa polypeptide through the membrane (hydrophobic signal peptide) and its insertion into the cell membrane (membrane anchorage hydrophobic sequence). It was also reported (3) that the insulin receptor polypeptide is related by its primary aa sequence to the tyrosine kinase family of oncogenes.

The development of methods to crosslink insulin to its receptor has shown that the α subunit is predominantly labeled by radioactive insulin derivatives (4). With the use of a radioactive and photoreactive insulin derivative [labeled in the side chain of lysine in position 29 in the insulin B chain (B29)] that is cleavable through an azo linkage, it was possible to transfer the radioactivity from the insulin B chain to the insulin receptor subunit. The polypeptide bands of 130 Kd (α subunit of the receptor) and 90 Kd (β subunit of the receptor) were labeled, the former more intensely. (5). Elastase cleavage of the 130 Kd subunit to low molecular weight peptides followed by HPLC separation revealed radioactivity in several short peptides (5).

The availability of the primary amino acid sequence of proinsulin, insulin receptor (3), insulin-like growth factors, as well as polypeptides coded by oncogenes of the tyrosine kinase family in computer data banks (6), allows for rapid computer analyses to reveal aa homologies that might help in the elucidation of functional domains in the polypeptides, as well as the mechanism of action of the insulin receptor (3). A computer program developed on the basis of the secondary structure predictions of Chou and Fasman (7) and Garnier *et al.* (8) is now available in the University of Wisconsin Genetic Computer group (UWGCG) software (6). With the aid of these programs, the secondary structure of proinsulin, insulin and the insulin receptor polypeptides was studied. It is also possible to compare both insulin and the insulin receptor polypeptide at the secondary structure conformational level.

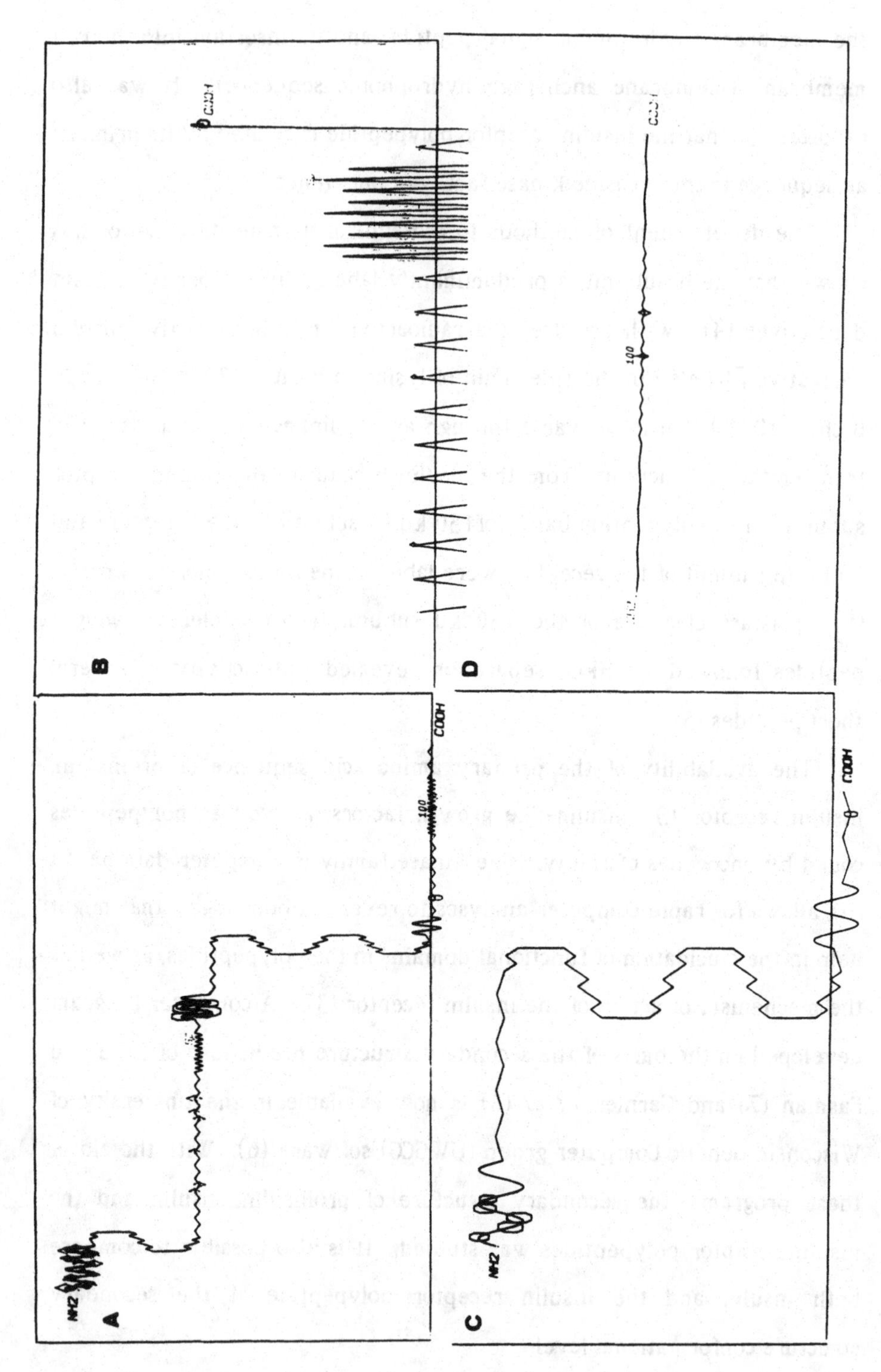
A
NH2
COOH
100
B
COOH
C
NH2
COOH
D
100

In this chapter, the secondary structure of human proinsulin and insulin was compared with that of pig insulin; the secondary structure of the insulin receptor was compared with that of the epidermal growth factor receptor and related oncogenes which were reported (3) to resemble the insulin receptor at the level of the primary aa sequence. Comparison by computer of the sequence of the insulin B chain and the insulin receptor revealed two sequences of partial homology in the INSR.

COMPUTER ANALYSIS OF POLYPEPTIDES

All the aa sequences [except those of the insulin receptor, which were obtained from Ullrich *et al.* (3)] were from the NBRF protein sequence data bank included in the software received from the UWGCG (6). The prediction of the secondary structure of the polypeptide was done with a computer program submitted to UWGCG by B.A. Jameson and H.Wolf (University of Munich, FRG). The program calculates the secondary structure of a peptide, using the Chou-Fasman (7) or Robson-Garnier (8) predictions and Kyte and Doolittle (9) hydropathic calculations. The secondary structure is graphically plotted, using the Gould Colorwriter.

The GAP program of the UWGCG package was used to produce optimal alignment between two sequences by inserting gaps in either one, as necessary. GAP uses the alignment method of Needleman and Wunsch (10). Comparison of two unrelated polypeptides by the GAP programs results in a random similarity of 10% or less.

Fig. 1. Chou-Fasman prediction of the secondary structure of human proinsulin (A), insulin B chain (aa 25-54) (B), insulin A chain (C) (aa 90-110) (D), and C peptide (aa 57-87). The alpha-helical sequence is displayed as ⋀⋀, turns as ⌣⌒⌣ and β sheets as —— or ⋀⋀⋀⋀. Hydrophilic aa are marked by an octagon and hydrophobic aa by a diamond (◇).

SECONDARY STRUCTURE OF PROINSULIN

The secondary structure of proinsulin calculated by the Chou-Fasman method is presented in Fig. 1A. The signal peptide comprises the first 24 aa at the N terminus of the polypeptide and contains a strongly hydrophobic α helical sequence of 13 hydrophobic aa. The sequence aa 25-54 is the B chain of the mature insulin molecule (Fig. 1B), and sequence aa 90-110 is the A chain of the mature insulin molecule (Fig. 1D). These peptides have a β-sheet conformation. However, the conformation of the insulin B chain (Fig. 1B) is different from that of the A chain. The C peptide (aa 57-87), which is released from the proinsulin molecule due to proteolytic cleavage, consists of α helical sequences flanked by turns (Fig. 1A and1C) and has the property of strong flexibility which may allow the proinsulin molecule to attain its tertiary structure. This, in turn, allows the development of S-S bonds between two cysteines in each of the A and B chains. It is of interest to note that the insulin B chain contains six hydrophobic aa, while the A chain has only two.

COMPARISON BETWEEN HUMAN AND PIG PROINSULINS

The primary sequence of human proinsulin (11-20) was compared by the GAP program to the sequence of pig proinsulin (21). Table I shows that the B chain of human insulin (aa 25-54) is identical to the B chain of pig insulin (aa 1-29), and the A chain of human insulin (aa 90-110) is almost identical to the A chain of pig insulin (aa 62-84). The difference is in aa 54 (T) in human proinsulin and aa 30 (A) in pig proinsulin. The signal peptide is missing from the pig proinsulin sequence. The difference between the human and pig proinsulins is expressed in an overall similarity of 88% and only 66% homology in the C peptide, as opposed to almost 100% homology in the A and B chains of the two proinsulins. As

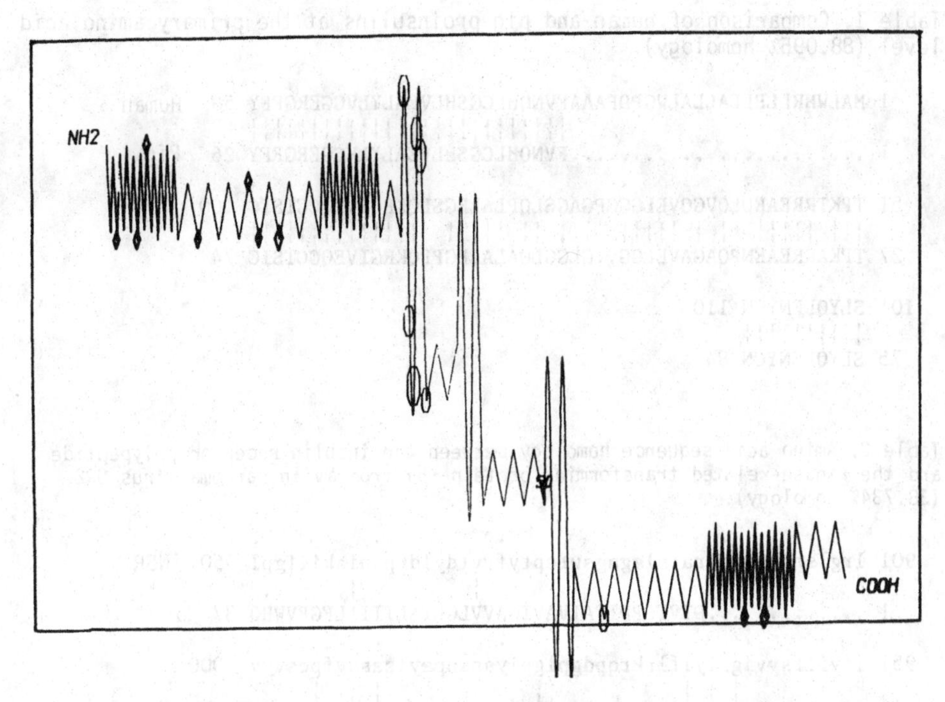

Fig. 2. Chou-Fasman prediction of the secondary structure of pig proinsulin. Symbols as in Fig. 1.

can be seen in Fig. 2, the C peptide of the pig proinsulin has three α helical sequences, as compared to two in the human C peptide of proinsulin (Fig. 1A and 1C). The B chain of the pig insulin contains one additional hydrophobic aa, as compared to the human B chain. The function of the hydrophobic aa in the B chain of insulin is not yet known.

THE SECONDARY STRUCTURE OF THE HUMAN INSULIN RECEPTOR POLYPEPTIDE

The 1370 aa sequence reported by Ullrich *et al.* (3) was analyzed by the computerized Chou-Fasman program and is displayed in Fig. 3. The molecule contains a hydrophobic domain at the N terminus that

Table 1. Comparison of human and pig proinsulins at the primary amino acid level (88.095% homology).

```
              .         .         .         .         .
   1 MALWMRLLPLLALLALWGPDPAAAFVNQHLCGSHLVEALYLVCGERGFFY 50   Human
                             ||||||||||||||||||||||||||
   1 ........................FVNQHLCGSHLVEALYLVCGERGFFY 26   Pig
              .         .         .         .         .
  51 TPKTRREAEDLQVGQVELGGGPGAGSLQPLALEGSLQKRGIVEQCCTSIC 100
     ||| |||||  | | |||||  | | || |||||  ||||||||||||||
  27 TPKARREAENPQAGAVELGG..GLGGLQALALEGPPQKRGIVEQCCTSIC 74
              .
 101 SLYQLENYCN 110
     ||||||||||
  75 SLYQLENYCN 84
```

Table 2. Amino acid sequence homology between the insulin receptor polypeptide and the kinase-related transforming protein *ros* from avian sarcoma virus UR2 (38.734% homology).

```
              .         .         .         .         .
 901 lrglspgnysvriratslagngswteptyfyvtdyldvpsniakiigpl 950  INSR
                                    |         |
   1 .............DTVTSPDITAIVAVIGAVVLGLTSLTIIILFGFVWHQ 37   ros
              .         .         .         .         .
 951 ifvflfsvvigsiylflrkrqpdgplgplyassnpeylsasdvfpcsvyv 1000
               | | |           |           |    |
  38 RWKSRKPASTGQIVLVKEDKELAQLRGMAETVGLANACYAVSTLPSQAEI 87
              .         .         .         .         .
1001 pdewevsrekitllrelgqgsfgmvyegnardi.ikgeaetrvavktvne 1049
            | |  |   || | || |||| | ||   |  | ||||||
  88 ESLPAFPRDKLNLHKLLGSGAFGEVYEGTALDILADGSGESRVAVKTLKR 137
              .         .         .         .         .
1050 saslrerieflneasvmkgftchhvvrllgvvskgqptlvvmelmahgdl 1099
      |   |  ||| ||  |  |   |   ||||     |     |||  |||
 138 GATDQEKSEFLKEAHLMSKFDHPHILKLLGVCLLNEPQYLILELMEGGDL 187
              .         .         .         .         .
1100 ksylrslrpeaennpgrppptlqemiqmaaeiadgmaylnakkfvhrdla 1149
      ||||  |         |  ||         |  |  ||    |||||||
 188 LSYLRGAR...KQKFQSPLLTLTDLLDICLDICKGCVYLEKMRFIHRDLA 234
              .         .         .         .         .
1150 arncmv......ahdftvkigdfgmtrdiyetdyyrkggkgllpvrwmap 1193
     |||| |           |||||||  ||||  ||||| | ||||||||||
 235 ARNCLVSEKQYGSCSRVVKIGDFGLARDIYKNDYYRKRGEGLLPVRWMAP 284
              .         .         .         .         .
1194 eslkdgvfttssdmwsfgvvlweitslaeqpyqglsneqvlkfvmdggyl 1243
     ||| |||||  || | |||  ||   |  ||| ||||  ||  |  || |
 285 ESLIDGVFTNHSDVWAFGVLVWETLTLGQQPYPGLSNIEVLHHVRSGGRL 334
              .         .         .         .         .
1244 dqpdncpervtdlmrmcwqfnpnmrptfleivnllkddlhpsfpevsffh 1293
       | |||    |||  ||   |  ||||  |   |    |       |
 335 ESPNNCPDDIRDLMTRCWAQDPHNRPTFFYIQHKLQEIRHSPLCFSYFLG 384
              .         .         .         .         .
1294 seenkapeseelemefedmenvpldrsshcqreeaggrdggsslgfkrsy 1343
       |  ||
 385 DKESVAPLRIQTAFFQPL................................ 402
```

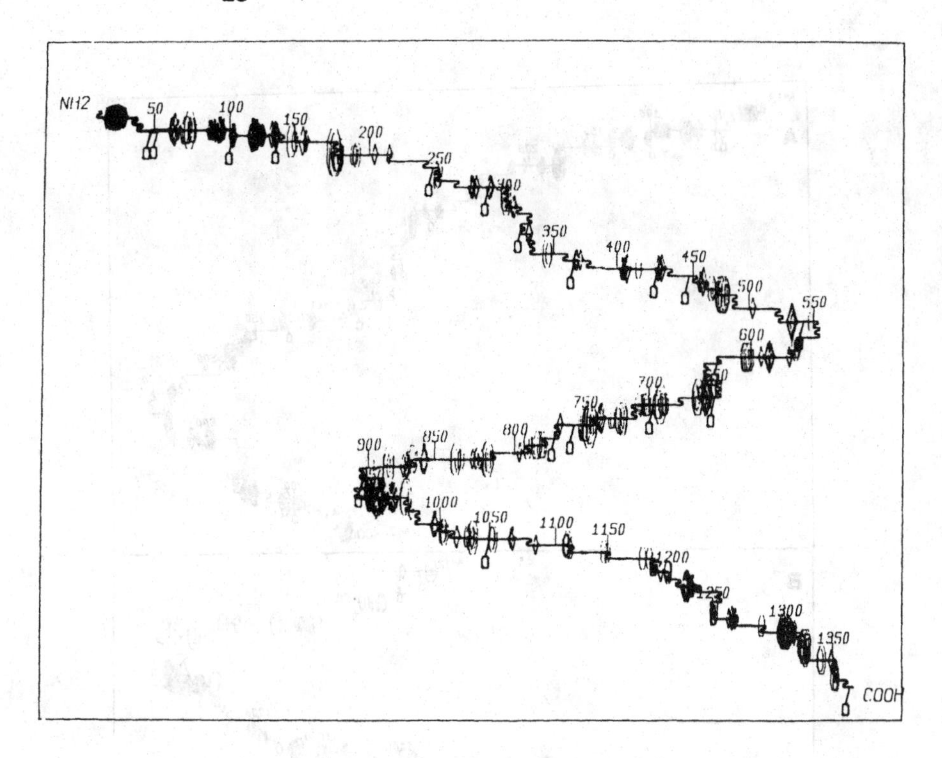

Fig. 3. Chou-Fasman prediction of the secondary structure of insulin receptor (based on the primary aa sequence in ref. 3). The symbols are as in Fig. 1. The boxes denote possible glycosylation sites. A hydrophobic signal peptide is located at the N terminus and a membrane anchorage domain is located at aa 940-970.

corresponds to the 27 aa signal peptide and an extended hydrophobic domain that corresponds to the membrane anchorage sequence reported by Ullrich *et al.* (3). The α chain of the insulin receptor (Fig. 4A) spans from aa 1 (aa 28 in Fig. 3, since the secondary structure analysis includes the signal peptide) to aa 724 (aa 751 in Fig. 3). The β subunit (Fig. 4B) is the rest of the insulin receptor polypeptide and contains the hydrophobic domain which serves as the membrane anchorage (aa 940-970).

Fig. 5 presents the secondary structure of the membrane anchorage domain of the insulin receptor which is present in the β subunit. The

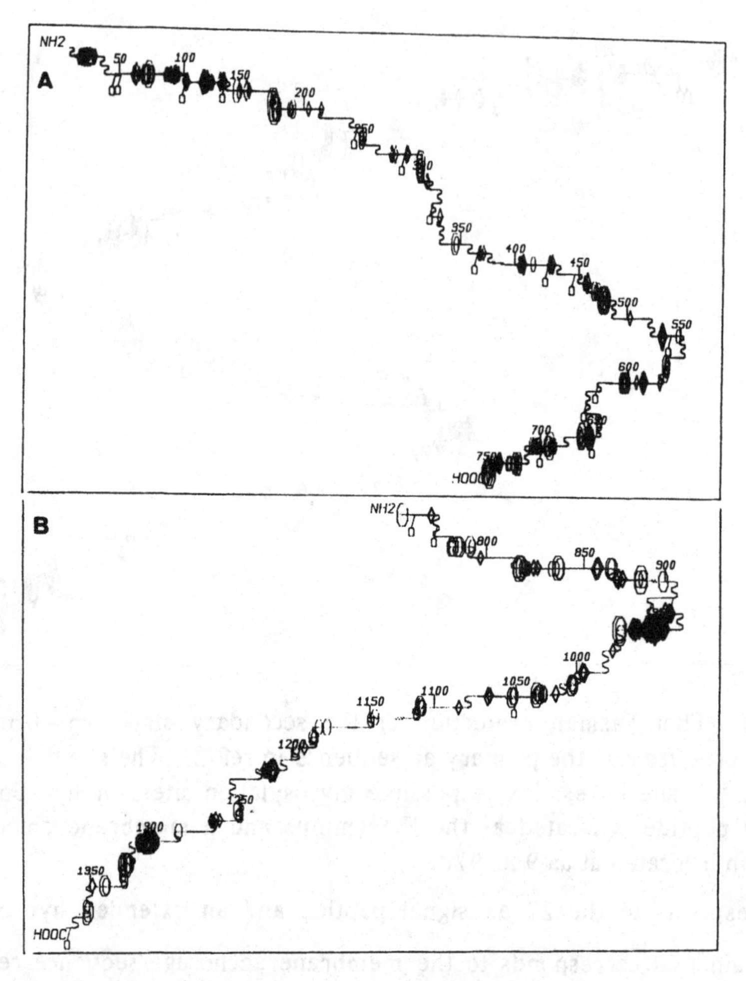

Fig. 4. Chou-Fasman secondary structure prediction of insulin receptor alpha subunit (A) and beta subunit (B). Symbols as in Fig. 1.

membrane anchorage domain has a conformation of a β sheet containing 17 hydrophobic amino acids. Both the α and β subunits are made of a series of β-sheets connected by flexible aa arranged in turns indicating the flexibility of both subunits. At least four short α helical sequences are present in the α subunit, while at least eight α helical domains are present in the β subunit. Seven of these α helical sequences contain hydrophilic aa. When the polypeptides attain their tertiary structure, the flexible

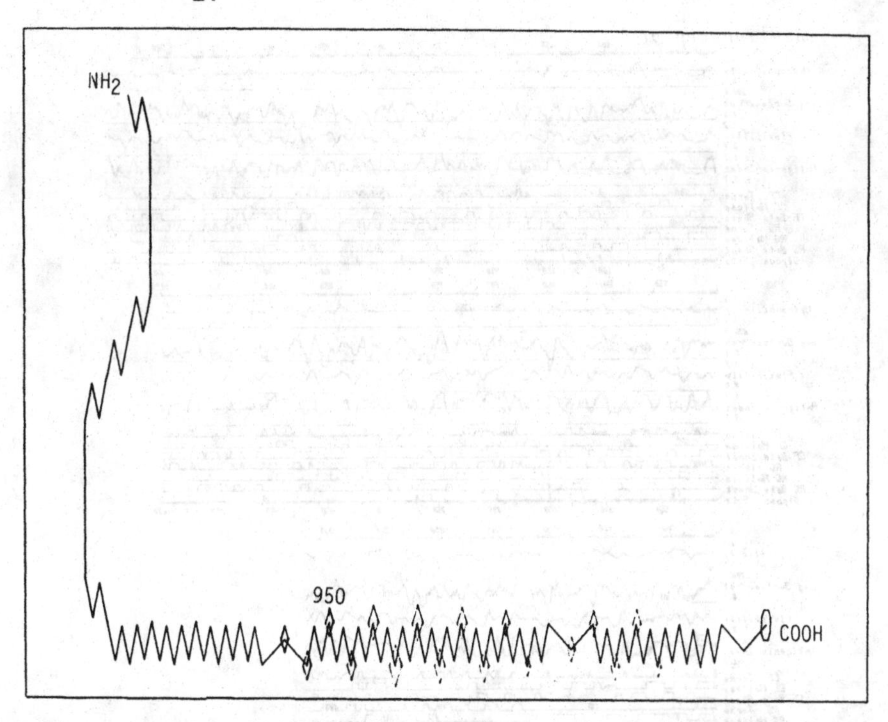

Fig. 5. Secondary structure of the membrane anchorage domain of the insulin receptor polypeptide (aa 940-970) with 17 hydrophobic aa. Symbols as in Fig. 1.

hydrophilic aa are arranged on the outside and the hydrophobic aa are inside the folded polypeptide. The full account of the number of α helical sequences is presented in Fig. 6, which summarizes the properties of the insulin receptor polypeptide calculated both by the Chou-Fasman and Robson-Garnier methods.

SECONDARY STRUCTURE COMPARISON BETWEEN INSULIN RECEPTOR POLYPEPTIDE AND THE TYROSINE KINASES PRODUCED BY ONCOGENES

Ullrich *et al.* (3) reported that the protein coded by the oncogene *ros* of avian sarcoma virus strain UR2 has a primary sequence homologous with the insulin receptor tyrosine kinase domain, rather than with all the

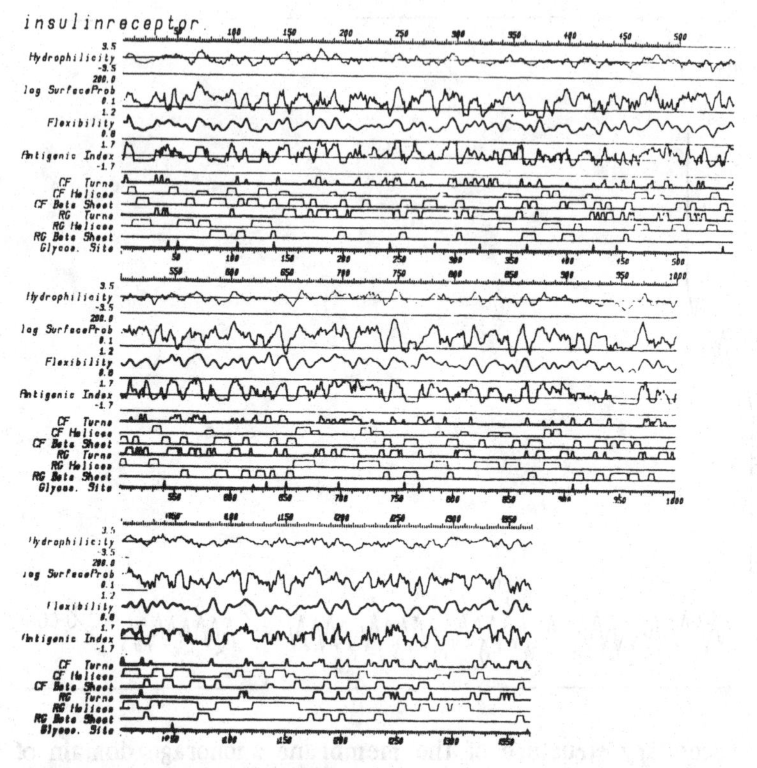

Fig. 6. Physical properties of the insulin receptor polypeptide. The analysis displays the conformational properties (turns, alpha helices and beta sheets) calculated according to the methods of Chou-Fasman (7) and Robson-Garnier (8). Surface probability and flexibility of the polypeptide are also presented. The program provides an assessment of the antigenic properties and the possible glycosylation sites in the polypeptide.

known *src* gene family members. Table 2 shows a 38.734% aa similarity between the sequence of the insulin receptor polypeptide and the *ros*-coded protein (22). The homology (using the GAP computer program of UWGCG) is between the *ros* aa sequence and the insulin receptor aa sequence spanning aa 932 (aa 905 in ref. 3, which excludes the 27 aa of the signal peptide) and aa 1300 (aa 1272, ref. 3) without the signal peptide, namely the intracytoplasmic domain of the insulin receptor. However, the tyrosine (ref. 4) in position aa 960 (aa 987 in Table 2) in the insulin receptor, which is a candidate for autophosphorylation, is not matched by a

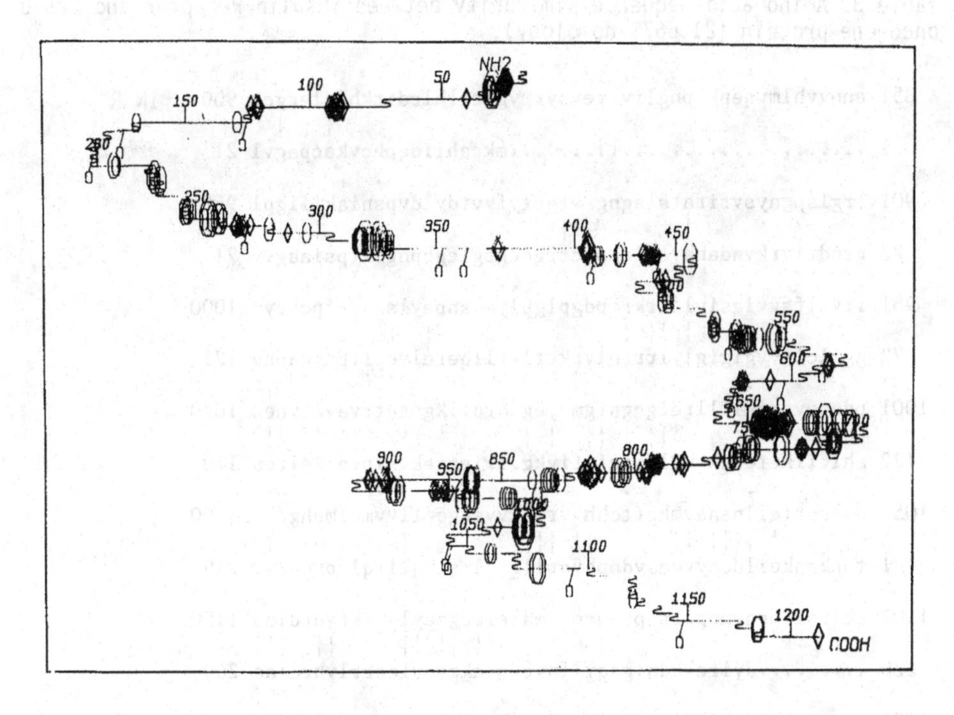

Fig. 7. Chou-Fasman prediction of the secondary structure of *erb* B oncogene-coded protein. Symbols as in Fig. 1.

tyrosine in the *ros* polypeptide. The tyrosine in position aa 1150 (aa 1177 in Table 2), which might also be a candidate for autophosphorylation (4), is indeed matched by a tyrosine in the *ros* polypeptide. In fact, the domain near these tyrosine residues [aa1161 (1202 in Table 2) in the insulin receptor and aa 252-293 in the *ros* protein] share a 82.926% homology.

In this respect, *ros* tyrosine kinase resembles the β subunit of the insulin receptor, but instead of a 200 aa extracytoplasmic domain of the β subunit, *ros* has a relatively short extracellular domain. Thus, the β subunit of the insulin receptor may have a tyrosine kinase activity like the *ros* oncogene protein, but the latter may lack a receptor function.

The insulin receptor polypeptide shows a 21.667% similarity to the

Table 3. Amino acid sequence similarity between insulin receptor and *erb* B oncogene protein (21.667% homology).

```
 851 ennvvhlmwqepkepnglivlyevsyrrygdeelhlcdtrkhfalergcr  900   INSR

   1 .............................mkcahfidgphcvkacpagvl   21   erb B

 901 lrglspgnysvriratslagngswteptyfyvtdyldvpsniakiiigpl  950

  22 gendtlvrkyadanavcqlchpnctrgckgpglegcpngsktpsiaagvv   71

 951 ifvflfsvvigsiylflrkrqpdgplgplyassnpeylsasdvfpcsvyv 1000

  72 ggllclvvvglgiglylrrrhivrkrtlrrllqerelvepltpsgeapnq  121

1001 pdewevsrekitllrelgqgsfgmvyegnardiikgeaetrvavktvnes 1050

 122 ahlrilketefkkvkvlgsgafgtiykg.lwipegekvkipvaikelrea  170

1051 aslrerieflneasvmkgftchhvvrllgvvskgqptlvvmelmahgdlk 1100

 171 tspkankeildeayvmasvdnphvcrllgicltstvqlitqlmpygcll.  219

1101 sylrslrpeaennpgrppptlqemiqmaaeiadgmaylnakkfvhrdlaa 1150

 220 .........dyirehkdnigsqyllnwcvqiakgmnyleerrlvhrdlaa  260

1151 rncmvahdftvkigdfgmtrdiyetdyyrkggkgllpvrwmapeslkdgv 1200

 261 rnvlvktpqhvkitdfglakllgadekeyhaeggkvpikwmalesilhri  310

1201 fttssdmwsfgvvlweitslaeqpyqglsneqvlkfvmdggyldqpdncp 1250

 311 ythqsdvwsygvtvwelmtfgskpydgipaseissvlekgerlpqppict  360

1251 ervtdlmrmcwqfnpnmrptfleivnllkddlhpsfpevsffhseenkap 1300

 361 idvymimvkcwmidadsrpkfreliaefskmardpprylviqgdermhlp  410

1301 eseelemefedmenvpldrsshcqreeaggrdggsslgfkrsyeehipyt 1350

 411 sptdskfyrtlmeeedmedivdadeylvphqgffnspstsrtpllsslsa  460

1351 hmnggkkngriltlprsnps.............................. 1370

 461 tsnnsatncidrngqghpvredsfvqryssdptgnfleesiddgflpape  510
```

erb B oncogene (reviewed in 23), as shown in Table 3. The 604 aa sequence of the *erb* B protein has the highest homology when aa1 of *erb* B protein is aligned with the insulin receptor sequence aa 880 (27 aa of the signal peptide included) and ends at aa 1370 of the insulin receptor, which is aligned to aa 480 of the *erb*B protein. A tyrosine is present in the *erb*B (aa 289) protein close to the tyrosine aa1177 in the insulin receptor. The membrane anchorage domains of the two polypeptides share homologous amino acids (aa 958-970 in insulin receptor and aa 79-91 in *erb*B protein).

The secondary structure of the *erb* B protein is presented in Fig. 7. The external portion of the *erb* B polypeptide has a relatively short and flexible sequence with many turns and lacks either hydrophilic or hydrophobic aa, in contrast to the external part of insulin receptor β subunit (Fig. 4B) which is longer and contains seven hydrophilic and four hydrophobic domains. The intracellular portion of *erb* B protein has numerous hydrophilic and hydrophobic domains, as does the intracellular domain of the insulin receptor β subunit.

COMPARISON OF INSULIN RECEPTOR WITH EPIDERMAL GROWTH FACTOR RECEPTOR

The *erb* B oncogene was reported (24) to share a marked homology with the primary aa sequence of the epidermal growth factor receptor (25, 26). Using the computer GAP program to compare the primary aa sequences of the two polypeptides, a 83.195% similarity between aa 580-1183 of epidermal growth factor receptor (EGFR) and the 604 aa sequence of the *erb* B oncogene protein was obtained (Table 4). The EGFR is also regarded like *erb* B protein to function as a tyrosine kinase (24). Fig. 8 displays the secondary structure of EGFR with the membrane anchorage domain near aa 650. In spite of the strong homology between *erb* B oncogene protein and EGFR, the latter shows only 13.313% similarity with the insulin receptor polypeptide.

The 21.6% similarity between *erb* B protein and insulin receptor β

Table 4. Amino acid sequence homology between *erb* B protein and epidermal growth factor receptor (EGFR) (83.195% homology).

```
              .         .         .         .         .
 551 ENSECIQCHPECLPQAMNITCTGRGPDNCIQCAHYIDGPHCVKTCPAGVM 600   EGFR
                                    ||| |||||||| |||||
   1 .............................mkcahfidgphcvkacpagvl 21    erb B
              .         .         .         .         .
 601 GENNTLVWKYADAGHVCHLCHPNCTYGCTGPGLEGCPTNGPKIPSIATGM 650
     ||| ||| |||||  || ||||||| || |||||||| || | |||| |
  22 gendtlvrkyadanavcqlchpnctrgckgpglegcp.ngsktpsiaagv 70
              .         .         .         .         .
 651 VGALLLLLVVALGIGLFMRRRHIVRKRTLRRLLQERELVEPLTPSGEAPN 700
     || || | || |||||  ||||||||||||||||||||||||||||||||
  71 vggllclvvvglgiglylrrrhivrkrtlrrllqerelvepltpsgeapn 120
              .         .         .         .         .
 701 QALLRILKETEFKKIKVLGSGAFGTVYKGLWIPEGEKVKIPVAIKELREA 750
     || ||||||||||| |||||||||| ||||||||||||||||||||||||
 121 qahlrilketefkkvkvlgsgafgtiykglwipegekvkipvaikelrea 170
              .         .         .         .         .
 751 TSPKANKEILDEAYVMASVDNPHVCRLLGICLTSTVQLITQLMPFGCLLD 800
     |||||||||||||||||||||||||||||||||||||||||||| |||||
 171 tspkankeildeayvmasvdnphvcrllgicltstvqlitqlmpygclld 220
              .         .         .         .         .
 801 YVREHKDNIGSQYLLNWCVQIAKGMNYLEDRRLVHRDLAARNVLVKTPQH 850
     | ||||||||||||||||||||||||||| ||||||||||||||||||||
 221 yirehkdnigsqyllnwcvqiakgmnyleerrlvhrdlaarnvlvktpqh 270
              .         .         .         .         .
 851 VKITDFGLAKLLGAEEKEYHAEGGKVPIKWMALESILHRIYTHQSDVWSY 900
     |||||||||||||| |||||||||||||||||||||||||||||||||||
 271 vkitdfglakllgadekeyhaeggkvpikwmalesilhriythqsdvwsy 320
              .         .         .         .         .
 901 GVTVWELMTFGSKPYDGIPASEISSILEKGERLPQPPICTIDVYMIMVKC 950
     ||||||||||||||||||||||||| ||||||||||||||||||||||||
 321 gvtvwelmtfgskpydgipaseissvlekgerlpqppictidvymimvkc 370
              .         .         .         .         .
 951 WMIDADSRPKFRELIIEFSKMARDPQRYLVIQGDERMHLPSPTDSNFYRA 1000
     ||||||||||||||| ||||||||| ||||||||||||||||||| |||
 371 wmidadsrpkfreliaefskmardpprylviqgdermhlpsptdskfyrt 420
              .         .         .         .         .
1001 LMDEEDMDDVVDADEYLIPQQGFFSSPSTSRTPLLSSLSATSNNSTVACI 1050
     || |||| | |||||| | |||| ||||||||||||||||||||   ||
 421 lmeeedmedivdadeylvphqgffnspstsrtpllsslsatsnnsatnci 470
              .         .         .         .         .
1051 DRNGLQSCPIKEDSFLQRYSSDPTGALTEDSIDDTFLPVPEYINQSVPKR 1100
     |||| |  |  |||| |||||||||   | |||| ||| ||| ||  ||
 471 drng.qghpvredsfvqryssdptgnfleesiddgflpapeyvnqlmpkk 519
              .         .         .         .         .
1101 PAGSVQNPVYHNQPLNPAPS...RDPHYQDPHSTAVGNPEYLNTVQPTCV 1147
     |          |     | |    |  ||  ||||| ||||||| |
 520 pstamvqnqiynfisltaisklpmdsryqnshstavdnpeylntnqspla 569
              .         .         .         .         .
1148 NSTFDSPAHWAQKGSHQISLDNPDYQQDFFPKEAKPNGIFKGSTAENAEY 1197
        | |   | | | ||| |||||||||| |
 570 ktvfesspywiqsgnhqinldnpdyqqdflptscs............... 604
```

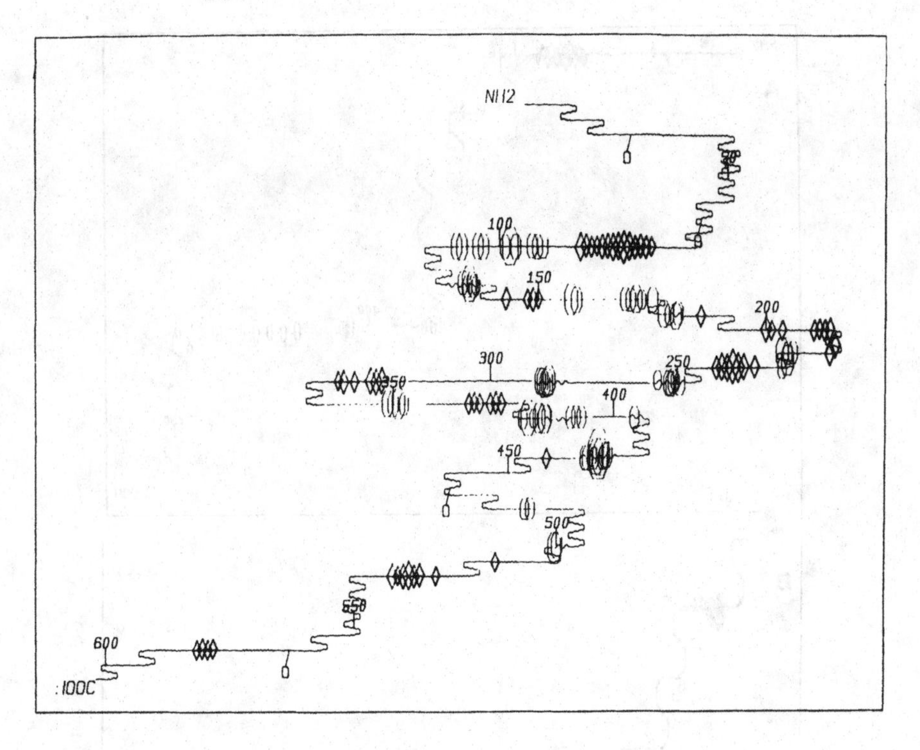

Fig. 8. Chou-Fasman prediction of the secondary structure of epidermal growth factor receptor (EGFR). Symbols as in Fig. 1.

subunit suggested a comparison also with EGFR polypeptide. Although the overall similarity between insulin receptor and EGFR is only 13.313%, there are domains along the two polypeptides with a stronger homology. Comparison of the membrane anchorage sequence in the two polypeptides (Fig. 8 - EGFR and Fig. 3 - insulin receptor) reveals that the domain is near aa 650 in EGFR, while in the insulin receptor, this domain is near aa 950. However, the carboxy termini of both polypeptides have the tyrosine kinase activity.

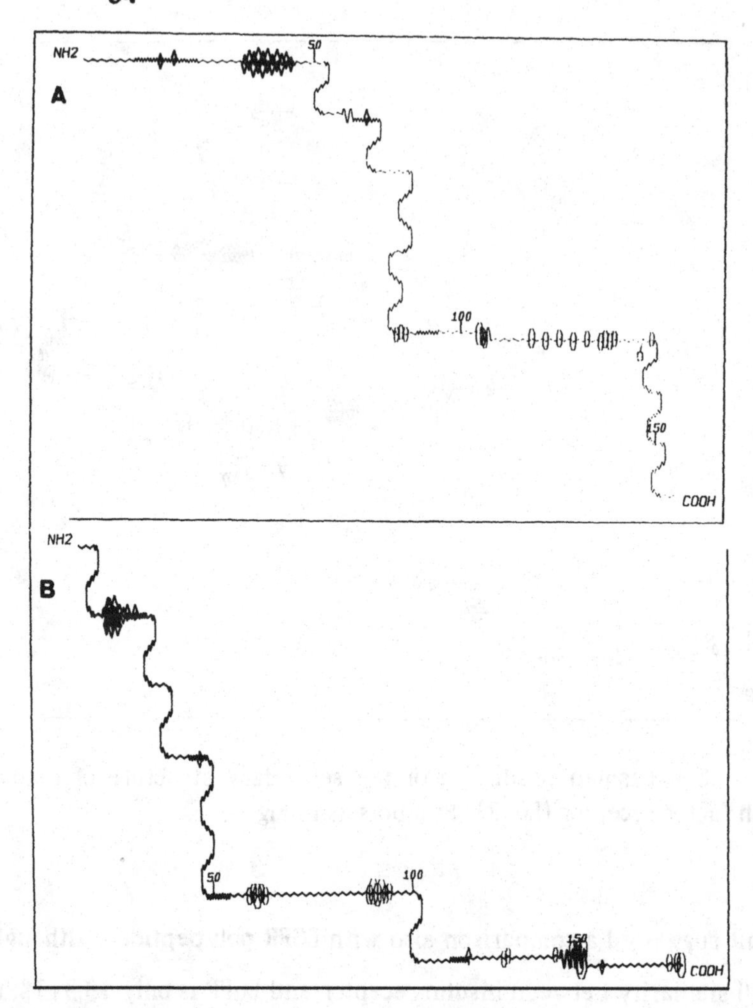

Fig. 9. Chou-Fasman prediction of the secondary structure of insulin growth factors I (A) and II (B). Symbols as in Fig. 1.

COMPARISON BETWEEN PROINSULIN AND INSULIN-LIKE GROWTH FACTORS 1 AND 2 (IGF1 and IGF2)

As shown by Bell *et al.* (1), the gene for IGF2 is situated in human chromosome 11 downstream from the insulin gene, while IGF1 is located on another chromosome. Since IGF1 and IGF2 have an insulin-like activity on cell growth, it was of interest to compare the secondary structure of proinsulin to that of IGF1 and IGF2. The secondary structure of IGF1 (Fig.

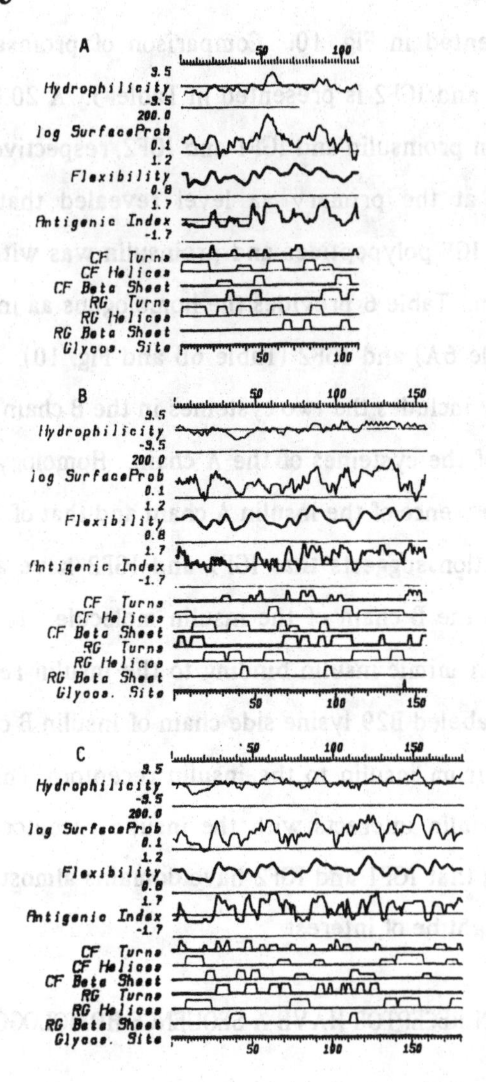

Fig. 10. Comparison of the calculated Chou-Fasman and Robson-Garnier predictions for the secondary structure of proinsulin (A), IGF1 (B) and IGF2 (C).

9A) and IGF2 (Fig. 9B) revealed a signal peptide at the N terminus of the molecule followed by flexible turns in both molecules with one β sheet (in IGF1) and a β sheet interrupted by a turn in IGF2.

Comparison of the secondary structure properties of proinsulin, IGF1 and IGF2, calculated according to Chou-Fasman and Robson-Garnier methods, is presented in Fig. 10. Comparison of proinsulin primary aa sequence to IGF1 and IGF2 is presented in Table 5. A 20.0% and 21.818% similarity between proinsulin and IGF1 and IGF2, respectively, were found. This comparison at the primary aa level revealed that the homology between the two IGF polypeptides and proinsulin was with a sequence in the insulin B chain. Table 6 provides the homologous aa in insulin B chain and in IGF1 (Table 6A) and IGF2 (Table 6B and Fig. 10). It is of interest that the homology includes the two cysteines in the B chain which form S-S bonds with two of the cysteines of the A chain. Homology was not found between the aa sequence of the insulin A chain and that of IGF1 or IGF2.

This observation suggests that IGF1 and IGF2 have a domain which closely resembles the B chain of the insulin molecule. It is possible that IGF1 and IGF2 can mimic insulin binding to the insulin receptor. Ng and Yip (5) used the labeled B29 lysine side chain of insulin B chain to transfer the radioactivity from insulin to the insulin receptor. This suggests that the B chain of insulin interacts with the insulin receptor. The computer analysis revealing that IGF1 and IGF2 have domains almost identical to the insulin B chain might be of interest.

DOES THE INSULIN RECEPTOR HAVE A SEQUENCE HOMOLOGOUS TO INSULIN B CHAIN?

In view of the strong possibility that the insulin B chain binds to the insulin receptor (5), it was of interest to determine if there are domains in the insulin receptor polypeptide that resemble the insulin B chain. As shown in Tables 5C, 5D, and Table 6, the insulin receptor contains a sequence homologous to the insulin B chain between aa 858-878 (in the β subunit) and aa 217-246 (in the α subunit). It is interesting that the homology includes two cysteines in either the α or β subunit of the insulin

Table 5. Comparison of amino acid sequences of proinsulin, insulin-like growth factors 1 and 2 and insulin receptor.

A. Proinsulin versus insulin-like growth factor 1 (20.0% similarity)

```
               .         .         .         .         .
   1  SEAMGKISSLPTQLFKCCFCDFLKVKMHTMSSSHLFYLALCLLTFTSSAT 50   IGF1
                                |            | |      |
   1  ..........................MALWMRLLPLLALLALWGPDPAAA 24   Proins.

               .         .         .         .         .
  51  AGPETLCGAELVDALQFVCGDRGFYFNKPTGYGSSSRRAPQTGIVDECCF 100  
           |||  || ||  ||| |||     |          |
  25  FVNQHLCGSHLVEALYLVCGERGFFYTPKTRREAEDLQVGQVELGGGPGA 74

               .         .         .         .         .
 101  RSCDLRRLEMYCAPLKPAKSARSVRAQRHTDMPKTQKEVHLKNASRGSAG 150
       |     ||
  75  GSLQPLALEGSLQKRGIVEQCCTSICSLYQLENYCN.............. 110
```

B. Proinsulin versus insulin-like growth factor 2 (21.818% similarity)

```
               .         .         .         .         .
   1  MGIPMGKSMLVLLTFLAFASCCIAAYRPSETLCGGELVDTLQFVCGDRGF 50   IGF2
               | ||  ||      ||      |||  ||  |  ||| |||
   1  ..MALWMRLLPLLALLALWGPDPAAAFVNQHLCGSHLVEALYLVCGERGF 48   Proins.

               .         .         .         .         .
  51  YFSRPASRVSRRSRGIVEECCFRSCDLALLETYCATPAKSERDVSTPPTV 100
             |          |         |              |    |
  49  FYTPKTRREAEDLQVGQVELGGGPGAGSLQPLALEGSLQKRGIVEQCCTS 98

               .         .         .         .         .
 101  LPDNFPRYPVGKFFQYDTWKQSTQRLRRGLPALLRARRGHVLAKELEAFR 150

  99  ICSLYQLENYCN...................................... 110
```

C. Insulin receptor compared to proinsulin (14.545% similarity)

```
               .         .         .         .         .
 851  ennvvhlmwqepkepnglivlyevsyrrygdeelhlcdtrkhfalergcr 900  INSR β
                    |                   |||      ||   |
   1  ......MALWMRLLPLLALLALWGPDPAAAFVNQHLCGSHLVEALYLVCG 44   Proins.
               .         .         .         .         .
 901  lrglspgnysvriratslagngswteptyfyvtdyldvpsniakiiigpl 950
       ||        |       |               |       |  |
  45  ERGFFYTPKTRREAEDLQVGQVELGGGPGAGSLQPLALEGSLQKRGIVEQ 94
               .         .         .         .         .
 951  ifvflfsvvigsiylflrkrqpdgplgplyassnpeylsasdvfpcsvyv 1000
            |      |
  95  CCTSICSLYQLENYCN.................................. 110
```

D. Insulin B chain compared to insulin receptor α subunit (23.333% similarity)

```
               .         .         .         .         .
  25  .......FVNQHLCGSHLVEALYLVCGERGFFYTPKT............. 54   Ins.
             |       | | |  |   |      |
 651  nithylvfwerqaedselfeldyclkglklpsrtwsppfesedsqkhnqs 700  INSR α
```

receptor which are positioned as in the insulin B chain (Table 6).

DISCUSSION

The availability of data banks of amino acid sequences of peptides and computer programs that calculate, predict and graphically display secondary structures of proteins made possible the analysis of the proinsulin and the insulin receptor polypeptide. It was possible to detect in each polypeptide the rigid (α helix and β sheet) and flexible (turn) regions and the distribution of hydrophobic and hydrophilic domains in each polypeptide molecule. The hydrophobic domains at the N terminus of the insulin receptor form the signal peptide sequence that is utilized in the transfer of the polypeptide across the cellular membrane. The insulin receptor has an additional hydrophobic domain which serves as the cell membrane anchorage aa domain and is present in the β subunit of the insulin receptor. In addition, numerous short hydrophobic domains are present both in the α and β subunits of the INSR, indicating that when the polypeptide attains its tertiary conformational structure they will be present internally, while the hydrophilic domains (mostly turns) will be on the outside of the folded polypeptide.

Numerous hydrophilic domains can be seen in the INSR polypeptide, while the proinsulin has a hydrophilic domain only in the cleavage product designated the C peptide. The latter has a hydrophilic α helical sequence at its N terminus, followed by turns and an α helix at the carboxy terminus. It was found (Table 1) that the secondary structure of the insulin C peptide of human origin differs in its primary and secondary structures from that of the pig C peptide (Figs. 1A and 2). Recently, it was reported (27) that the C peptide of proinsulin specifically binds to NEDH rat pancreatic beta islet-cell-derived tumorogenic cells which secrete insulin. A homologous

Table 6. Homologous amino acid sequences in proinsulin, insulin-like growth factors 1 and 2 and insulin receptor.

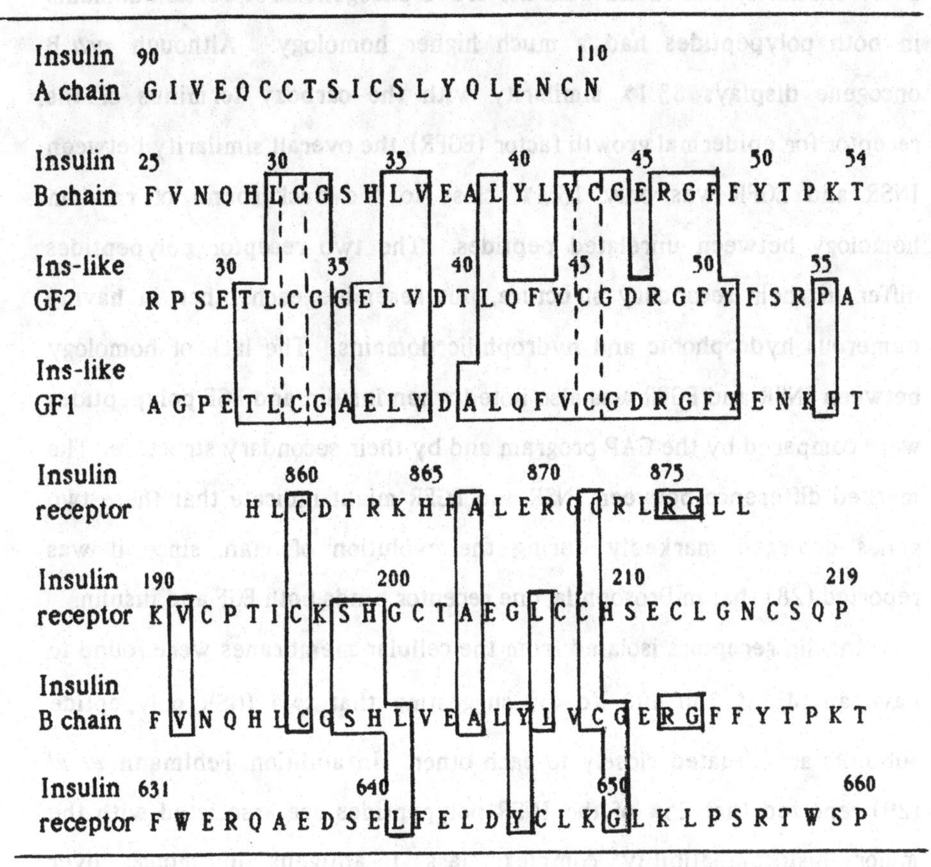

Insulin 90 110
A chain G I V E Q C C T S I C S L Y Q L E N C N

Insulin 25 30 35 40 45 50 54
B chain F V N Q H L G G S H L V E A L Y L V C G E R G F F Y T P K T

Ins-like 30 35 40 45 50 55
GF2 R P S E T L C G G E L V D T L Q F V C G D R G F Y F S R P A

Ins-like
GF1 A G P E T L C G A E L V D A L Q F V C G D R G F Y E N K P T

Insulin 860 865 870 875
receptor H L C D T R K H F A L E R G C R L R G L L

Insulin 190 200 210 219
receptor K V C P T I C K S H G C T A E G L C C H S E C L G N C S Q P

Insulin
B chain F V N Q H L C G S H L V E A L Y L V C G E R G F F Y T P K T

Insulin 631 640 650 660
receptor F W E R Q A E D S E L F E L D Y C L K G L K L P S R T W S P

synthetic C peptide inhibited glucose-stimulated insulin release from beta cells *in vivo* and *in vitro*. Thus, the C peptide might serve as a regulatory peptide for glucose metabolism.

The exact binding site for insulin on the INSR is not fully understood (C.C. Yip, personal communication). The studies by Ng and Yip (5) revealed that the radioactivity in insulin chain B29 lysine can be transferred to at least three positions in the INSR polypeptide.

Ullrich *et al.* (3) compared the primary aa sequence of INSR to that of the tyrosine kinase family of oncogenes. Indeed, a 38% similarity was found between the β subunit of the INSR and the *ros* oncogene, but only 21.6% similarity was found with the *erb* B oncogene. Yet, certain domains in both polypeptides had a much higher homology. Although *erb* B oncogene displays 83.1% similarity with the carboxy terminus of the receptor for epidermal growth factor (EGFR), the overall similarity between INSR and EGFR was only 13.3%, close to the background of random homology between unrelated peptides. The two receptor polypeptides differ in their secondary structure, but resemble each other in having numerous hydrophobic and hydrophilic domains. The lack of homology between INSR and EGFR was also noted when insulin and EGF polypeptides were compared by the GAP program and by their secondary structure. The marked difference between INSR and EGFR might indicate that these two genes diverged markedly during the evolution of man, since it was reported (28) that in Drosophila, one receptor binds both EGF and insulin.

Insulin receptors isolated from the cellular membranes were found to have an M_r of 350-400 Kd (3) suggesting that two INSR polypeptide subunits are situated closely to each other. In addition, Fehlmann *et al.* (29) reported that 25% of the INSR polypeptides are associated with the major histocompatibility complex class I antigens in mouse liver membranes. Insulin binding to the INSR activates the receptor's tyrosine kinase (30) due to autophosphorylation of a tyrosine aa 960 (987, including the signal peptide) or aa 1150 (1177, including the signal peptide 27 aa) (31). It is possible that conformational changes involving class I histocompatibility complex might affect activation of the tyrosine kinase of the INSR to phosphorylate the histocompatibility polypeptide. This might serve as a signal to the cell nucleus via its contact with the INSR. The β subunit of INSR has an ATP binding site (32, 33). The insulin activation of the INSR protein kinase was found to correlate with the histone kinase activity in the cell nucleus (30). It is of interest that phorbol esters which were found to cause breakage of chromosomal DNA in treated cells were also found to modulate INSR phosphorylation in cultured hepatoma cells (34).

Future studies on the binding of insulin to the insulin receptor will allow analysis of the mode of function of the insulin receptor in relation to its secondary and tertiary structures.

ACKNOWLEDGMENTS

The author wishes to thank Dr. N. Kaiser, Hadassah Medical Center, Jerusalem, for most helpful discussions. Special thanks are due to Prof. H.Wolf (University of Munich, FGR) for making his computer program available through UWGCG software. The study was supported by the Foundation for the Study of Molecular Virology and Cell Biology, Phoenix, Arizona, USA. The support of Mr. and Mrs. Harold A. Haytin, Beverly Hills, California, USA, is gratefully acknowledged.

REFERENCES

1. Xiang, K., Sanz N., Karam, J.H. and Bell, G.I. Chapter 1, this volume.
2. Grzeschik, K.-H. and Kazazian, H.H. Cytogenet. Cell Genet. 40:179-205, 1985.
3. Ullrich, A., Bell, J.R., Chen, E.Y., Herrera, R., Petruzzelli, L.M., Dull, T.J., Gray, A., Coussens, L., Liao, Y.-L., Tsubokawa, M., Mason, A., Seeburg, P.H., Grunfeld, C., Rosen, O.M. and Ramachandran, J. Nature 313:756-761, 1985.
4. Siegel, T., Ganguly, S., Jacobs, S., Rosen, O.M. and Rubin, C.S. J. Biol. Chem. 256:9266-9273, 1981.
5. Ng, D.S. and Yip, C.C. Biochem. Biophys. Res. Commun. 133:154-160, 1985.
6. Devereux, J., Haeberlin, P. and Smithies, O. Nucl. Acids Res. 12:387-395, 1984.
7. Chou, P.Y. and.Fasman, G.D. Adv. Enzymol. 47:45-147, 1978.
8. Garnier, J., Osguthorpe D.J. and Robson, B. J. Mol. Biol. 120:97-120, 1978.

9. Kyte, J. and Doolittle, R.J. J. Mol. Biol. 157:105-132, 1982.
10. Needleman, S.B. and Wunsch, C.D. J. Mol. Biol. 48:443-453, 1970.
11. Bell, G.I., Pictet, R.L., Rutter, W.J., Cordell, B., Tischer, E. and Goodman, H.M. Nature 284:26-32, 1980.
12. Ullrich, A., Dull, T.J., Gray, A., Brosius, J. and Sures, I. Science 209:612-615, 1980.
13. Bell, G.I., Swain, W.F., Pictet, R., Cordell, B., Goodman, H.M., and Rutter, W.J. Nature 282:525-527, 1979.
14. Sures, I., Goeddel, D.V., Gray, A., and Ullrich, A. Science 208:57-59, 1980.
15. Nicol, D.S.H.W. and Smith, L.F. Nature 187:483-485, 1960.
16. Oyer, P.E., Cho, S., Peterson, J.D. and Steiner, D.F. J. Biol. Chem. 246:1375-1386, 1971.
17. Ko, A., Smyth, D.G., Markussen, J., and Sundby, F. Eur. J. Biochem. 20:190-199, 1971.
18. Sieber, P., Kamber, B., Hartmann, A., Joehl, A., Riniker, B., and Rittel, W. Helv. Chim. Acta 57:2617-2621, 1974.
19. Naithani, V.K. Hoppe-Seyler's Z. Physiol. Chem. 354:659-672, 1973.
20. Geiger, R., Jaeger, G., and Koenig, W. Chem. Ber. 106:2347-2352, 1973.
21. Blundell, T., Dodson, G., Hodgkin, D., and Mercola, D. Adv. Protein Chem. 26:279-402, 1972.
22. Neckameyer, W.S. and Wang, L.H. J. Virol. 53:879-884, 1985.
23. Graf, T. and Beng, H. Cell 34:7-9, 1983.
24. Downward, J., Yarden, Y., Mayes, E., Scrace, G., Totty, N., Stockwell, P., Ullrich, A., Schlessinger, J. and Waterfield, M.D. Nature 307:521-527, 1984.
25. Ullrich, A., Coussens, L., Hayflick, J.S., Dull, T.J., Gray, A. *et al.* (14 authors) Nature 309:270-273, 1984.
26. Mroczkowski, B., Mosig, G., and Cohen, S. Nature 309:270-273, 1984.
27. Flatt, P.R., Swanson-Flatt, S., Hampton, S.M., Bailey, C.J. and Marks, V. Bioscience Rep. 6:193-199, 1986.
28. Thompson, K.L., Decker, S.J. and Rosner, M. R. Proc. Natl. Acad. Sci. USA 82:8443-8447, 1985.
29. Fehlmann, M., Peyron, J.-F., Samson, M., Van Obberghen, E., Brandenburg, D. and Brossette, N. Proc. Natl. Acad. Sci. USA 82:8634-8637, 1985.
30. Klein, H.H., Friedenberg, G.R., Kladde, M. and Olefsky, J.M. J. Biol. Chem. 261:4691-4697, 1986.
31. Yu, K.T. and Czech, M.P. J. Biol. Chem. 261:4715-4722, 1986.
32. Shia, M.A. and Pilch, P.F. Biochemistry 22:717-721, 1983.
33. Van Obberghen, E., Ross, B., Kowalsky, A., Gazzano, H. and Ponzio, G. Proc. Natl. Acad. Sci. USA 80:945-949, 1983.
34. Takayama, S. *et al.* Proc. Natl. Acad. Sci. USA 81:7797-7801, 1984.

3

MONOCLONAL ANTIBODIES TO PANCREATIC LANGERHANS ISLETS: IMMUNOCHEMISTRY AND APPLICATIONS

K. KRISCH, S. SRIKANTA, G. HORVAT, I. KRISCH, R. C. NAYAK, J. POSILLICO, P. BUXBAUM, N. NEUHOLD, O. BRAUN

Department of Pathology and Second Department of Surgery, University of Vienna, Medical School, Vienna, Austria; Joslin Diabetes Center, Brigham and Women's Hospital, Harvard Medical School, Boston, MA, USA

ABSTRACT

Monoclonal islet cell antibodies have begun to facilitate the identification and biochemical characterization of cellular differentiation molecules. We have developed a series of murine monoclonal antibody probes directed towards islet endocrine cells. Utilizing these antibodies we identified and biochemically characterized several islet cell antigens. Many islet cell antibodies appear to react with unique carbohydrate residues on their antigen molecules. As demonstrated by immunocytochemistry, islet cells share several monoclonal antibody-defined antigens and antigenic determinants with other neuroendocrine cell types indicating common modes of functional differentiation and specialization.

Some applications of the monoclonal islet cell antibodies in diagnostic pathology as markers for neuroendocrine tumors as well as for the analysis of the specific biologic function of monoclonal antibody-defined antigens illustrate the usefulness of monoclonal antibodies as probes to elucidate complex molecular mechanisms involved in normal and pathological endocrine cell function.

INTRODUCTION

Over the last decade it has become increasingly clear that Type I diabetes mellitus (DM) results from a slowly progressive, immunologically mediated, islet beta cell destruction, initiated in genetically predisposed individuals by as yet undefined (environmental or non-genetic) factors (1-8).

Becker, Y (ed), Virus Infections and Diabetes Melitus. © 1987 Martinus Nijhoff Publishing, Boston. ISBN 0-89838-970-4.

A prolonged asymptomatic preclinical stage of progressive beta cell destruction associated with specific immunological abnormalities (islet cell antibodies, activated T cells) precedes the onset of overt Type I DM by several years (2-5). Autoimmune mechanisms being involved in the beta cell destructive process in Type I DM are indicated by several lines of evidence: 1) the demonstration of lymphocytic infiltration of the islets ("insulitis") early in the course of the disease, 2) clinical association of Type I DM with other established or putative autoimmune diseases (both endocrine and non-endocrine), 3) the increased prevalence of organ-specific autoantibodies (thyroid, gastric parietal cells, adrenal gland), 4) the detection of a family of islet-cell antibodies (islet cell cytoplasmic antibody - ICA, islet cell cytoplasmic complement fixing antibody - CF-ICA, islet cell surface antibody - ICSA, and islet cell cytotoxic antibody), 5) demonstration of abnormalities of cell-mediated immune function, 6) association with certain antigens of the major histocompatibility locus, and 7) studies in animal models of Type I DM (for review see 9 and 10; 11-15).

Despite these recent advances in our knowledge, the basic mechanism of the autoimmune beta cell destruction remains poorly understood. The selectivity of the immune response which leads to the beta cell destruction in Type I DM appears to be conferred by unique differentiation antigens ("target" autoantigens) expressed by the pancreatic islet beta cells. The autoantibody detected by indirect immunofluorescence staining of a normal human pancreatic frozen section by sera of patients with newly diagnosed Type I DM has been conventionally referred to as the cytoplasmic islet cell antibody (ICA), based on its presumptive binding to antigens located within the cytoplasm of the islet cells (14). These autoantibodies selectively bind to all the endocrine pancreatic cells, but not to the exocrine acinar pancreatic cells, ductular cells or the stromal connective tissue cells. Anti-islet autoantibodies have also been detected by the surface staining of viable human and rodent pancreatic islet cells, and these have been referred to as islet cell surface antibodies (ICSA; 15). These autoantibodies directed against pancreatic islet cells constitute important predictive and pathogenetic hallmarks of Type I DM (2, 3, 5, 19-21). Serum or immunoglobulin fractions containing islet-cell autoantibodies suppress glucose-induced insulin release (22). Islet-cell antibodies exert this effect by binding to their corresponding target antigen(s) on the islet-cell surface. This suggésts that, besides their role in autoimmunity, these autoantigens may have a fundamental physiological role as receptors ("glucose sensor"?) or key regulatory molecules of endocrine pancreatic islet cells. Besides their binding to circulating islet-cell autoantibodies, these autoantigens may also be important in the generation of the cell-mediated anti-islet autoimmune response (9, 10). The evidence for autoreactive T-lymphocytes is provided by 1) the production of migration-inhibition factor and lymphocyte blast transformation in response to islet antigens, 2) cell-mediated cytotoxicity against insulinoma cells, and 3) in vitro inhibition of insulin release by T-lymphocytes from Type I diabetic subjects. Also, during the course of "insulitis", mononuclear cells infiltrate (presumably attracted by the beta-cell autoantigens), and selectively destroy the beta cells of the intact pancreatic islets as well as those at foci of beta-cell regeneration. Despite their central role in the autoimmunity and beta cell destruction of Type I DM, the biochemical nature and the mechanism of action of the target islet-cell autoantigens is currently unclear (except for a single report of a 64 kD islet cell protein, immunoprecipitated by a few diabetic sera (16)).The polyclonality and low titers of circulating islet-cell antibodies have precluded further detailed analysis of the immunochemistry of islet-cell autoantigens.

Due to their unlimited availability and defined specifity, monoclonal antibodies (Mabs; 17) represent powerful probes for investigating the biochemical nature and function of specialized tissue and cell-specific differentation antigens. Successful applications of these techniques to endocrine research include the production of Mabs to endocrine cell surface differentiation antigens, studies of immunoendocrinology to define cell surface target antigens and the human "organspecific" autoantibodies to define abnormalities of immunoregulation in diseases such as Graves' disease and DM (for review see 18).

In view of the paucity of available information concerning the biology and molecular nature of islet-cell antigens, over the last 5 years, utilizing the hybridoma technique (17), we have created a library of monoclonal islet-cell antibodies. These Mabs together with those generated by other investigators have permitted the preliminary biochemical characterization and functional evaluation of several novel protein-, glycoprotein-, and glycolipid-islet cell antigens (23-42).

For this review we will describe the methodology used for the generation of murine monoclonal islet antibodies, the tissue distribution and preliminary biochemical characterization of their target antigens, and illustrate with specific examples successful applications of these Mab probes to cellular biology and diagnostic pathology.

GENERATION OF MONOCLONAL ISLET CELL ANTIBODIES

Before the advent of Mabs it could be argued that a serological approach to identification of cellular antigens had several distinct advantages over conventional biochemical methodology. With the availability of Mabs these advantages have become overwhelming in comparison with any other single approach. The preparation of Mabs does not require pure antigens, and this is one of the most powerful aspects of the hybridoma method. The principle of generating Mabs is relatively simple. Briefly, the splenocytes of an appropriately immunized mouse (most commonly) are fused with a myeloma cell line, thus producing a hybrid cell, which retains the transformed phenotype of the parental myeloma cell (i.e.

it is effectively immortalized), as well as the specific antibody producing program of the splenic B lymphocyte. Such hybridomas are subject to single cell cloning and screening for production of antibody of the desired specificity. These selected cells, derived from a single cell producing antibody are monoclonal and hence produce Mabs.

Our techniques that have been successful in creating monoclonal islet cell antibodies are essentially the same as described by Scearce and Eisenbarth (28). Six-week-old female Balb/c mice were immunized by intraperitoneal injections on day 0, 7 and 14 with human islet cells isolated from cadaveric human pancreatic specimens (HISL-series) by enzymatic digestion and gradient centrifugation (42), bovine pancreatic islet cells (BISL-series), and rat pancreatic islet cells or RIN cells (RISL-series), suspended first in complete, and subsequently in incomplete Freund's adjuvant. The final injection did not contain adjuvant and splenocytes were fused on day 17 with the myeloma cell line P3X63-Ag8 or P3X63-Ag8.653 with the use of the standard polyethylene glycol technique (28, 31). The resulting hybrids were initially screened for their reactivity with human, bovine and rat pancreatic islets respectively with the use of an indirect immunofluorescence technique on cryostat sections. Positive colonies were cloned several times by limiting dilution and grown as ascites tumors in Balb/c mice that had been injected intraperitoneally with pristane 7 days prior to injection of the hybrids.

Mabs directed against pancreatic islet cells have also been generated by other approaches such as immortalization of autoantibody-producing B lymphocytes from humans (24) and animals with autoimmune diabetes (38-40). In addition, Mabs initially generated against non-islet antigen preparations have been shown to react with pancreatic islets during routine or systematic immunohistochemical screening (44, 45).

IMMUNOCYTOCHEMISTRY OF ISLET CELL ANTIGENS

The immunoreactivity of Mabs was evaluated by an indirect immunoperoxidase technique. Acetone-fixed frozen tissue sections were incubated with the Mabs (dilution 1:1000-1:4000 of ascites fluid) and subsequently incubated with peroxidase-conjugated rabbit anti-mouse Ig antibody (dilution 1:100) and then with peroxidase-conjugated swine anti-rabbit Ig antibody (dilution 1:100). Peroxidase activity was detected by the 3,3'-diaminobenzidine tetrahydrochloride (DAB) reaction, and the sections were counterstained with hematoxilin.

Table 1. Immunocytochemistry of islet cell antigens (recognized by HISL Mabs)

Tissue	HISL-1	HISL-4	HISL-5	HISL-7	HISL-8	HISL-9	HISL-14	HISL-19	HISL-22
Pancr. Islet	+++	+++	+++	+++	+++	+++	+++	+++	+++
Pancr. Acini	–	+	+	+ +	–	+	+	–	–
Pituitary, ant.	+++	+++	+++	++	+++	+++	+++	+++	+
post.	+++	–	–	++	+++	–	–	–	+++
Adrenal med.	+++	+++	+++	+	+++	+++	+++	+++	+++
cortex.	–	–	–	–	–	–	–	–	–
Thymus epithel.	–	+++	+++	++	–	+++	+++	+	–
Thymocytes	++	–	–	–	++	–	–	–	–
Parathyroid	+	–	+++	+	+	+++	+++	–	–
Thyroid Foll.	+	–	+	+	+	+	+	–	+
C cells	+++	+	+	–	++	+	+	+++	–
Gut epithel.	/	/	+	/	/	++	++	–	/
endocrine	/	/	+	/	/	+++	+++	+++	/
Liver Par.	–	++	++	+	–	++	++	–	+
Kidney Glom.	++	–	+	+++	++	+	+	–	++
Tub.	–	++	+++	+++	++	+++	+++	–	++

Table 1 and 2 summarize the immunohistochemical reactivity spectrum of monoclonal islet cell reactive antibodies on a variety of human neuroendocrine and non-neuroendocrine tissues. As noted in the table, none of the Mabs raised against pancreatic islets was islet-cell specific but cross reacted with other neuroendocrine or non-neuroendocrine tissues (Figs. 1-6). This cross reactivity may have developmental and functional significance and may have some implications in the co-occurrence of multiple autoimmune disorders

Table 2. Immunocytochemistry of islet cell antigens (recognized by Mabs BISL-26, BISL-32, A2B5, 3G5, and 4F2)

Tissue	BISL-26	BISL-32	A2B5	3G5	4F2
Pancr. Islet	A cells	A cells	+++	+++	+++
Pancr. Acini	–	–	–	–	–
Pituitary, ant.	+	/	+	+++	+
post.	–	/	+	–	+++
Adrenal med.	/	/	+	+	+
cortex.	/	/	–	–	++
Thymus epithel.	–	/	+	–	/
Thymocytes	/	/	–	+	/
Parathyroid	/	/	+	+	+++
Thyroid Foll.	–	/	/	–	++
C cells	/	/	+	/	++
Gut epithel.	/	/	–	/	/
endocrine	/	/	/	/	/
Liver Par.	/	/	–	–	–
Kidney Glom.	/	/	–	–	–
Tub.	/	/	+	–	+++

involving these cell types. For instance, besides reactivity with all cells of the pancreatic islet, Mab HISL-19 also reacted with other cells of the diffuse neuroendocrine system (DNES, 47), including anterior pituitary cells, C cells of the thyroid (Fig. 6b), endocrine cells of the gut (Fig. 5b) and bronchus, the adrenal medulla, and central (Fig. 6a) and peripheral neurons. The cross reactivity of islet cell antibodies with other neuroendocrine cells illustrates the well-known interrelationship between the cells of the DNES with regard to common modes of functional differentiation and specialization.

With a few exceptions (bronchial mucosa, gallbladder mucosa (Fig. 6c), small ducts of salivary glands), Mab HISL-19 did not react with other non-neuroendocrine tissues tested. It seems noteworthy that the reactivity of a single Mab with different cell types may also reflect identical or cross-reactive epitopes on structurally and functionally unrelated antigens and therefore further biochemical studies have to be performed in order to exclude this possibility (see below).

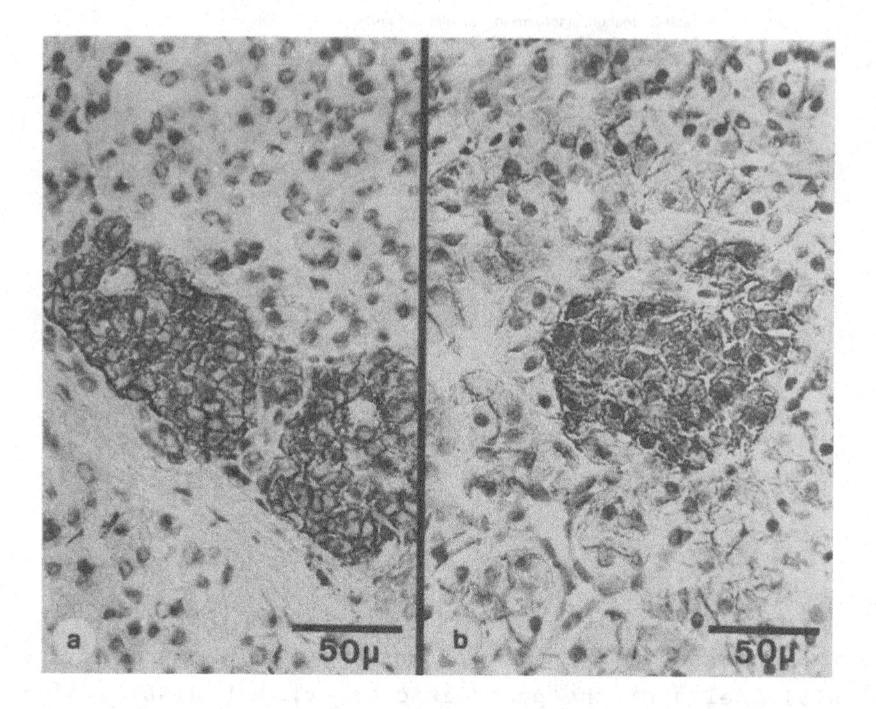

Fig. 1. Immunoperoxidase staining of pancreatic islets (human) with Mabs HISL-1 (a), and HISL-7 (b). Mab HISL-1 reacted with islet cells only, Mab HISL-7 reacted also but weakly with the acinar tissue (frozen tissue sections, counterstained with hematoxilin).

The species specificity of the Mabs was tested using human, bovine, and rat pancreatic tissues. All Mabs of the HISL-group except Mab HISL-19, which also reacted with bovine tissue, and HISL-22 which reacted also with bovine and rat tissue, reacted with human islet cells only. BISL-26, -32 and -37 Mabs reacted with bovine and human islet cells. However, in contrast to their reactivity with all cells of the bovine pancreatic islet cells, BISL-26 and -32 immunostained only the peripheral mantle (probably A cells) of the human pancreatic islet (Fig. 4a). BISL-30 and -31 were specific for bovine islet cells. Rat tissues were not tested in this group. Mab A2B5 and 3G5 and RISL-37 reacted with human, bovine and rat pancreatic islets, Mab 4F2 was human specific.

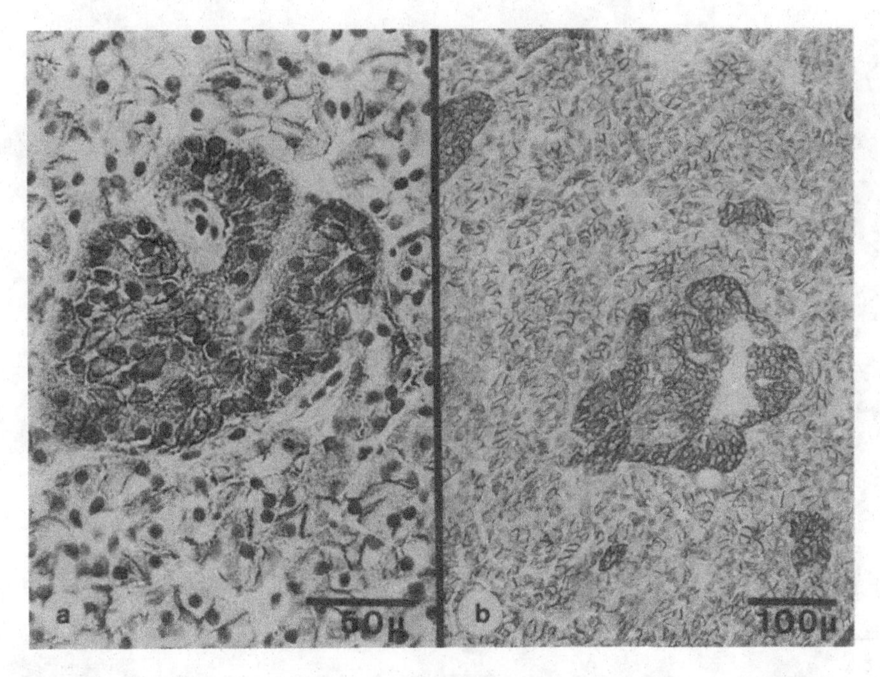

Fig. 2. Immunoperoxidase staining of pancreatic islets (human) with Mabs HISL-5 (a), and HISL-14 (b). Both Mabs reacted strongly with all cells of the pancreatic islet and weakly with acinar pancreatic cells (frozen tissue sections, counter-stained with hematoxilin).

The cellular localization of the corresponding antigenic determinants was evaluated using indirect immunofluorescence staining of isolated, viable islet cells. The antigens detected by Mabs of the HISL-group as well as the antigens recognized by antibody A2B5, 3G5, and 4F2 were located on the cell surface (except antibody HISL-19 (questionable result) and HISL-22, which was cytoplasmic), BISL-26, -30, -31, and -32 and RISL-39 all reacted with cytoplasmic antigenic determinants of islet cells. The latter observation is based on their lack of binding to the surface of isolated viable islet cells, despite their reactivity with intracellular islet structural components on frozen pancreatic sections.

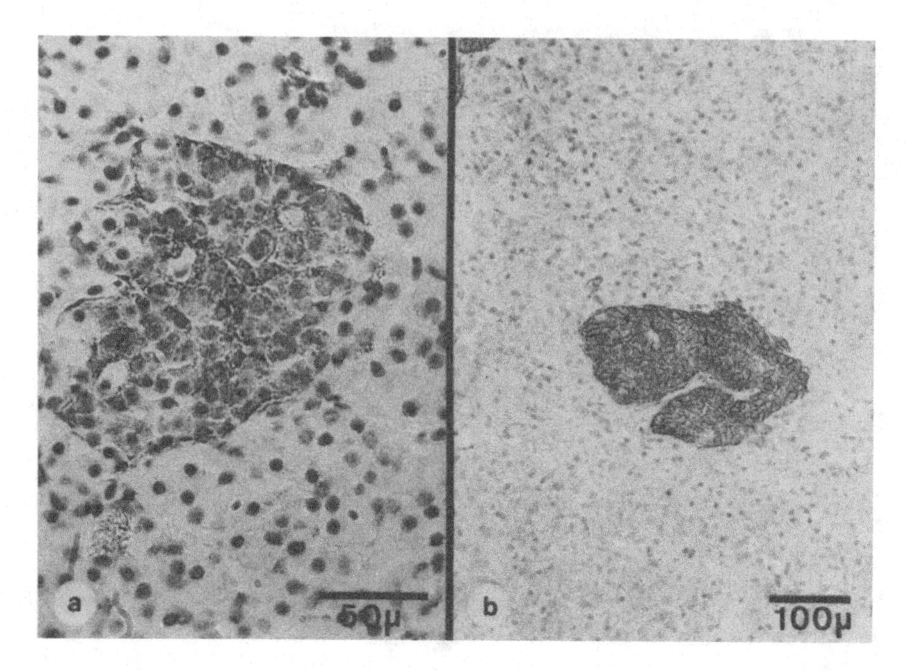

Fig. 3. Immunoperoxidase staining of pancreatic islets (human) with Mabs HISL-22 (a) and 4F2 (b). Only the islets are stained (frozen tissue sections, counterstained with hematoxilin).

BIOCHEMICAL CHARACTERIZATION OF ISLET CELL ANTIGENS

Protein and glycoprotein islet cell antigens:

Mabs HISL-9 and HISL-14 react with two different epitopes of a 100 kD surface glycoprotein as demonstrated by immunoaffinity purification from non-ionic detergent extracts of radiolabelled cells (lactoperoxidase-catalized radioiodination method for selective labeling of cell surface proteins (46)) using the corresponding Mab immunosorbent columns followed by sodium-dodecyl sulfate polyacrylamide gel electrophoresis (SDS-PAGE) and autoradiography (29).

As demonstrated by the "Western" immunoblotting technique using crude NP-40 extracts Mab HISL-19 recognized a series of four human islet cell proteins (p120, p69, p67, and p56) in one islet cell tumor (immunocytochemically and clinically non-

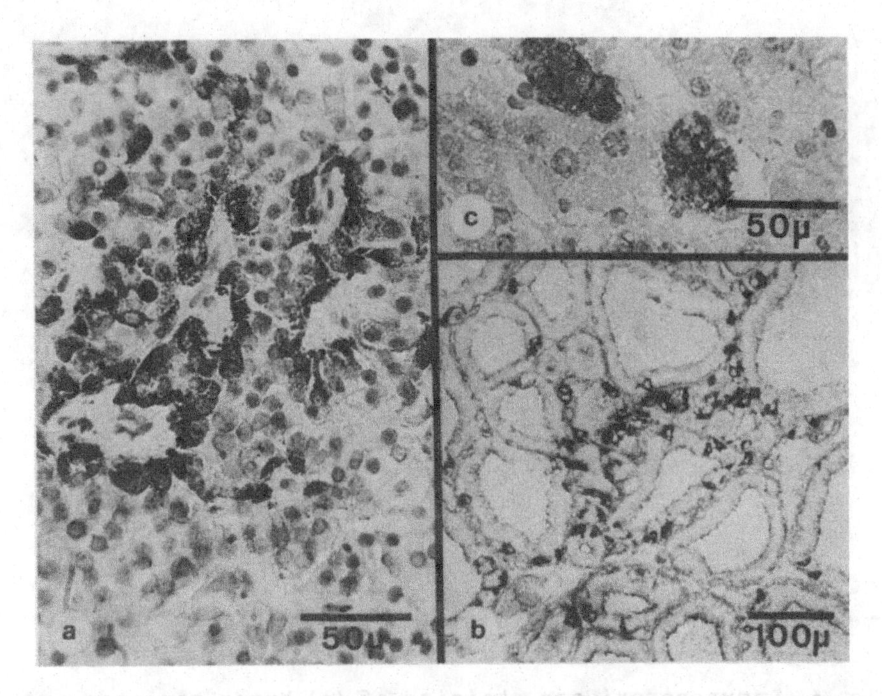

Fig. 4. a. Immunoperoxidase staining of pancreatic sections (human) with Mab BISL-26. In contrast to its reactivity with all cells of the bovine pancreatic islet (not shown), Mab BISL-26 immunostained only the peripheral mantle (A cells) of the human pancreatic islet. b. Immunostaining of (bovine) thyroid with Mab BISL-26. Mab BISL-26 reacted strongly with the thyroid C cells. Thyroid follicular cells demonstrated only very weak immunoreactivity restricted to the luminal cell surface with this antibody. c. Immunoperoxidase staining of human adrenal gland with Mab HISL-5. Mab HISL-5 reacted only with the cells of the adrenal medulla but not at all with the adrenal cortex (frozen sections, counterstained with hematoxilin).

functioning islet cell tumor, IH9; Fig. 7), and one bronchial carcinoid (32). Another islet cell tumor (gastrinoma, IH13) expressed the HISL-19 reactive p120, p69, and p67 species (data not shown), a third islet cell tumor (immunocytochemically and clinically non-functioning, IH14) expressed only the p 67 species (Fig. 8 and 9). The latter also bound to concanavalin A lectin, thus indicating its glycoprotein nature

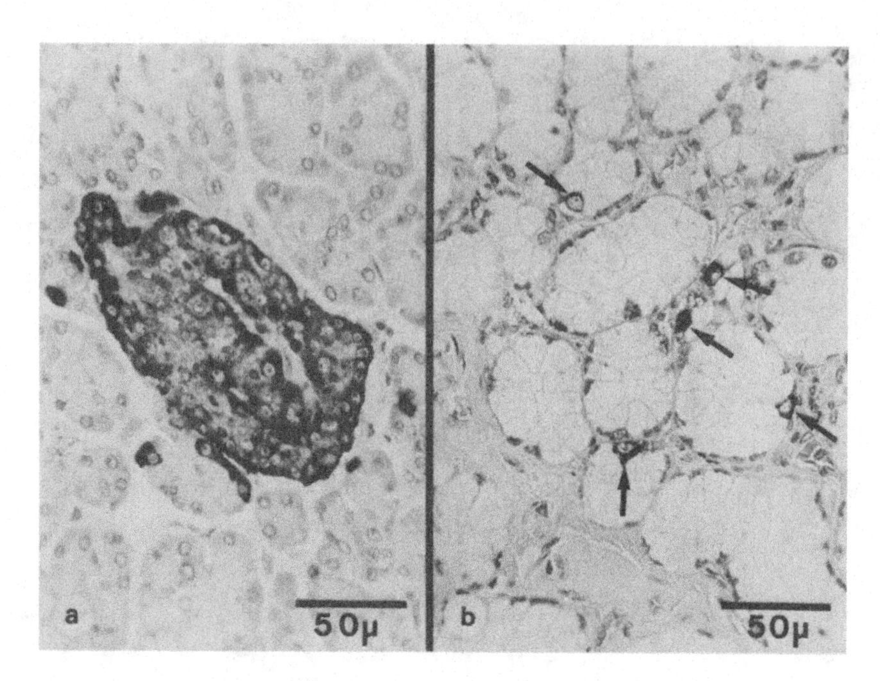

Fig. 5. Immunoperoxidase staining of (a) human pancreas and (b) human stomach (antrum) with Mab HISL-19, demonstrating its selective reactivity with pancreatic islets and gut endocrine cells (arrows) respectively (formalin fixed and paraffin embedded tissues, sections counterstained with hematoxilin).

(Fig. 8). The p120 species has also been identified in a medullary carcinoma and bovine islets. In the human pituitary gland, Mab HISL-19 recognized a different set of proteins with molecular weights of 40 and 24 kD (32).

So far the interrelationship between the neuroendocrine proteins detected by Mab HISL-19 is unknown. It is quite possible that Mab HISL-19 reacts with the same epitope on functionally unrelated antigens. However, the possibility of identical epitopes on variable sized products resulting from divergent post-translational cleavage and processing from a common precursor protein cannot be excluded.

Mab HISL-19 immunoaffinity chromatography of a NP-40 insuloma extract (IH14) under nonreducing conditions yielded a

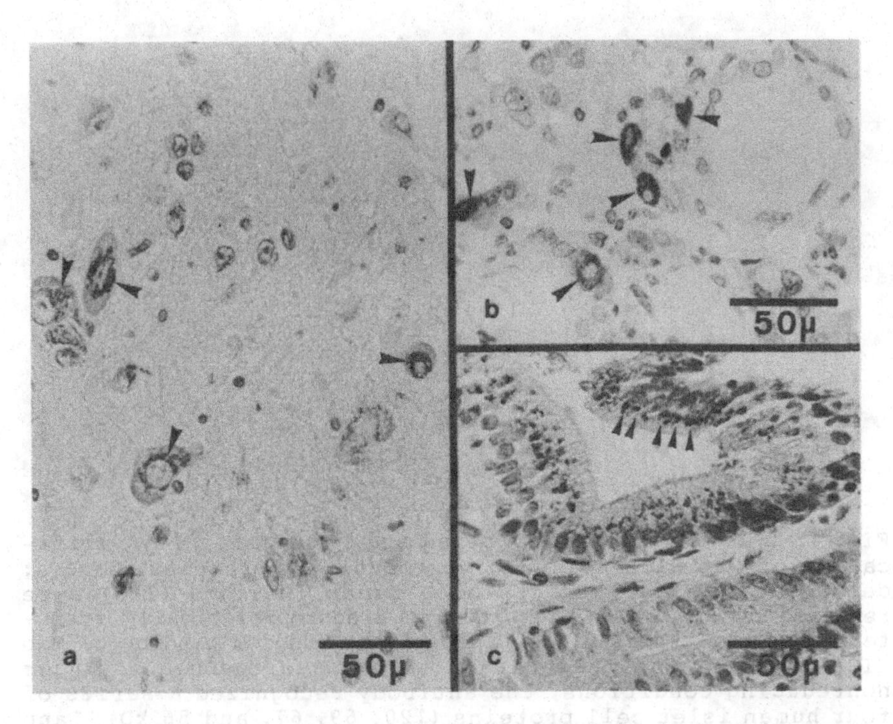

Fig. 6. Immunoperoxidase staining of (a) central neurons (hypothalamus) resulting in a ring-shaped reaction pattern (arrowheads) surrounding the cell nucleus, and (b) thyroid C cells (arrowheads) with Mab HISL-19 (human tissues). c. Mab HISL-19 immunoreactivity in gallbladder mucosa showing supranuclear cytoplasmic structures (arrowheads; formalin fixed and paraffin embedded tissues, counterstained with hematoxilin).

singular specific band of a m.w. equal to 67 kD in this case, as demonstrated by SDS-PAGE and "Western" immunoblotting (48; Fig. 9). In the presence of DTT, this protein could be separated into two apparantly identical subunits of a m.w. of 35 kD each. However, the possibility that the 67 kD protein is composed of only one 35 kD subunit bearing the epitope recognized by Mab HISL-19 and other low molecular weight subunits unrecognized by the immunoblot technique, cannot be excluded. Physical staining of the affinity purified HISL-19 antigen in SDS polyacrylamide gels with Coomassie Blue failed to give

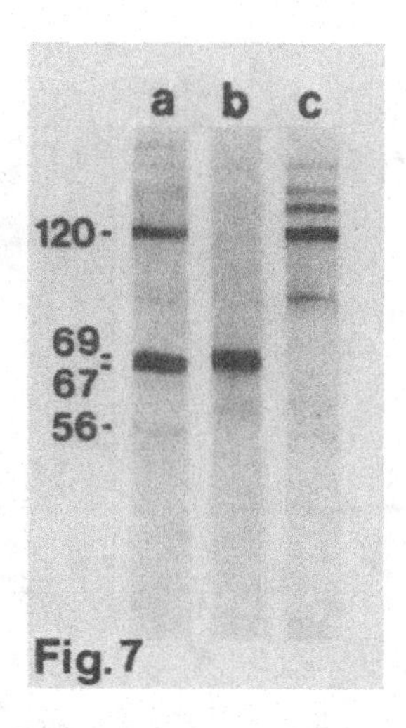

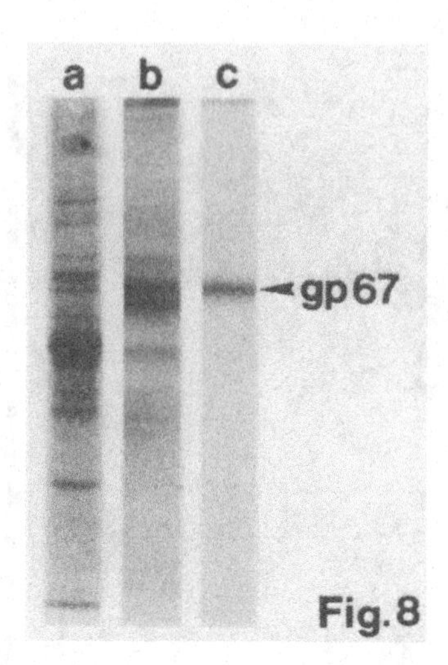

Fig. 7. "Western" immunoblotting and immunoaffinity purification of Mab HISL-19 reactive islet cell proteins. Lane a: detergent (NP-40) extracts of a human insuloma (IH9) were resolved by SDS-PAGE, the proteins electrophoretically transferred onto nitrocellulose paper and probed for binding to Mab HISL-19 utilizing an indirect peroxidase technique. Under nonreducing conditions, the antibody recognized a series of four human islet cell proteins (120, 69, 67, and 56 kD). Lane b and c: differences in affinity properties of the HISL-19 cell proteins (i.e. p120 versus p67-69) has allowed further separation of these molecular species during Mab HISL-19 immunoaffinity chromatography. The p67-69 species could be eluted from the Mab HISL-19 immunosorbent column by 0.1M glycine-HCl, pH 2.5 (lane b), the p120 protein required 3.5M potassium thiocyanate for elution (lane c).

Fig. 8. Lane a: profile of total cellular proteins of a human islet cell tumor (IH14) after SDS-PAGE and electrophoretical transfer onto nitrocellulose paper of a NP-40 extract stained with amido black. Lane b: overlays with biotinylated concanavalin A lectin using the Avidin-Biotin-Complex (ABC) method, demonstrated a major band in a molecular weight range of 67 kD. This protein was identified as the Mab HISL-19 reactive protein by immunooverlays with Mab HISL-19 (lane c).

any positive result, perhaps due to low amount of protein present.

Immunoblotting experiments using crude NP-40 extracts from human gallbladder mucosa (which reacted with Mab HISL-19 by immunocytochemistry) failed to reveal any protein band reacting with Mab HISL-19. However, enrichment of the HISL-19 antigen from NP-40 extracts of human gallbladder mucosa by Mab HISL-19 affintiy chromatography resulted in the identification of a protein band with identical m.w. characteristics as described above for the antigen detected by Mab HISL-19 in the insuloma (48; Fig. 9). These findings demonstrate, that the 67 kD protein similar to "neuron specific" enolase (NSE, 49) may not be considered as strictly neuroendocrine specific, but is expressed in a greater amount by neuroendocrine cells than by some non-neuroendocrine epithelia.

Preabsorption of Mab HISL-19 with the NSE protein and the anti-NSE antibody with the Mab HISL-19 reactive protein (p67) respectively, previous to immunostaining, did not abolish their reactivity with sections of islet cell tumors in both instances (48). On the other hand, the reactivity of these antibodies could be blocked after absorption with their corresponding antigens. These findings, as well as the different tissue distribution and different molecular weights (Fig. 9) distinguish the Mab HISL-19 reactive protein from NSE. Despite the fact, that chromogranin (50; in contrast to the HISL-19 protein) has never been demonstrated neither cytochemically nor biochemically to be present in the gallbladder epithelium, it seems unlikely, that the protein detected by Mab HISL-19 belongs to the chromogranin family. Recent results based on similar experiments as performed for the differentiation of the HISL-19 antigen from NSE using a Mab to chromogranin support this presumption (unpublished observations).

Mabs 4F2 (initially generated following immunization of Balb/c mice with human T cells (HSB-2 T cell line; 51)) and LC7-2 (generated following immunization of Balb/c mice with Pan-1 cell line derived from a ductal adenocarcinoma of the pancreas; provided by Drs. Vito Quaranta and Steve Levine, Scripp's Clinic, California) react with epitopes (heavy and light chain respectively) on a cell surface heterodimeric glycoprotein (approximate molecular weights 80:40 kD; 52). The 4F2/LC7-2 antigen is of particular interest in terms of the selected group of cells expressing antigens of identical

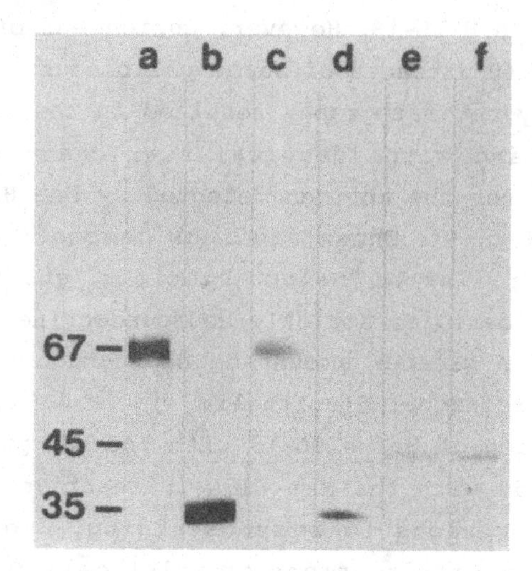

Fig. 9. "Western" immunoblotting after SDS-PAGE of Mab HISL-19 immunoaffinity chromatographed proteins of a human islet cell tumor (IH14; lane a and b) and human gallbladder mucosa (lane c and d) under nonreducing (lane a and c) and reducing (lane b and d) conditions. Lane e and f: reactivity of islet cell proteins (NP-40 insuloma extract) to polyclonal anti-NSE antibody under nonreducing (lane e) and reducing (lane f) conditions using the peroxidase-anti-peroxidase complex (PAP) technique.

subunit structure. Expression of the 4F2/LC7-2 antigen appears to be associated with cell proliferation and endocrine function. All malignant neoplasms and cell lines tested thus far, as well as many other rapidly proliferating cells express the 4F2/LC7-2 antigen. Endocrine cells expressing the 4F2/LC7-2 antigen include pancreatic islets, parathyroid cells and thyroid cells. The precise function of this cytosolic calcium regulatory (see below) cell membrane protein is currently unknown.

Glycolipid islet cell antigens:

Islet cells express glycolipid surface membrane molecules, and these are defined by Mabs A2B5 (44), 3G5 (27), and R2D6 (37). Mab A2B5, initially generated following immunization of Balb/c mice with chicken retina, reacts with a glycolipid of neurons with the solubility and chromatographic properties of a GQ ganglioside (53). Studies of the antigen expressed by RINm5f rat insulinoma cells have shown that the antigen is resistant to trypsin but sensitive to neuraminidase digestion, can be pelleted from a water homogenate, dissolved in chloroform/methanol and resuspended in water, strongly indicating its glycolipid nature (44).

Mab 3G5 was generated from Balb/c mice immunized with rat brain and reacts with a ganglioside common to neurons and pancreatic islets (27) but distinct from that of A2B5 as determined by tissue distribution. Antibody 3G5, adsorbed to polyvenyl plates, can immobilize islet ganglioside micelles containing several complex gangliosides. This ability has led to a solid phase radioassay to detect antiganglioside antibodies. When monoclonal anti-islet antibodies are produced, a number may react with complex gangliosides, including a recently described rat B cell specific Mab (37). The assay should provide a useful screening test for this class of monoclonals.

Because of limiting amounts of material, we were unable to characterize precisely the antigens detected by the other monoclonal islet cell antibodies. Preliminary results obtained by the "Western" immunoblotting procedure utilizing a human thyroid medullary carcinoma substrate are shown in table 3. The antigens recognized by Mabs HISL-4, -8, -22 and BISL-32 were not detected by the immunoblotting technique.

Table 3. Thyroid C cell proteins defined by monoclonal islet cell antibodies

Mab	Antigen
HISL - 1	p135
HISL - 5	p93
HISL - 7	p28
HISL - 9	p100
HISL - 14	p100
HISL - 19	p120

MONOCLONAL ANTIBODIES AS PROBES FOR ENDOCRINE CELL BIOLOGY AND PATHOPHYSIOLOGY

Mabs which have been generated to a diverse spectrum of endocrine cell differentiation molecules have served as powerful probes for defining novel cell surface antigens and their specific biologic functions. It is not surprising, therefore, that these reagents have proven to be exquisite analytical tools in essentially all fields of biological research including endocrinology.

Rassi et al (54) have shown that Mab A2B5, which reacts with a GQ ganglioside, inhibited (80-90%) insulin secretion stimulated by glucose, tolbutamide and N6 monobutyryl cyclic AMP. This inhibition was dose-dependent (0.67-67 μg/ml Mab). Another islet cell surface Mab, 3A4, has also been shown to inhibit significantly glucose-stimulated insulin release with or without complement present (55). In studies utilizing anti-glycolipid Mabs A2B5 and 3G5, binding of these Mabs to viable dispersed parathyroid cells resulted in a marked reduction (40-50%) of low calcium (0.75 mM) stimulated parathyroid hormone (PTH) release compared to controls (56). Moreover, hormone secretion by Mab-treated cells was similar to that observed from untreated cells suppressed in high extracellular calcium (1.5 mM). Mab BISL-37, which recognizes an intracellular antigen in the parathyroid cells, had no effect on PTH secretion. All together, these findings suggest a role for complex plasma membrane glycolipids in the regulation of transmembrane signalling/signal transduction and hormone biosynthesis/secretion in parathyroid cells, possibly through their interaction with extracellular secretogouges.

Binding of Mab 4F2 (heavy chain epitope) to human adenomatous dispersed parathyroid cells resulted in a marked (53.8$\pm$ 7.9%) reduction in low calcium - stimulated parathyroid hormone secretion to levels equivalent to cells suppressed by high extracellular calcium (Fig. 10a). Typically these functional effects were optimal at antibody dilutions of 1:10000 - 1:100000 ascites fluid. Parallel studies, using the calcium

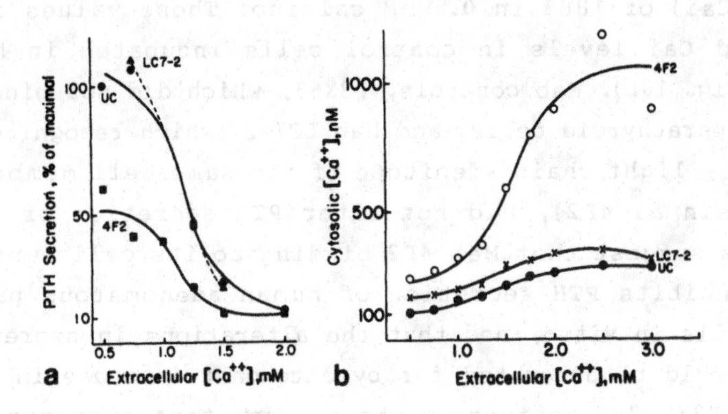

Fig. 10. A representative experiment illustrating the parathyroid secretory response and cytosolic calcium levels in Mab 4F2 treated parathyroid cells compared to controls (i.e. untreated cells (UC) or Mab LC7-2 treated cells). a. Parathyroid cells were treated with Mabs 4F2 or LC7-2 and incubated at varying extracellular calcium concentrations in order to determine the effects of these Mabs on hormone release. PTH secretion is represented as % of maximal and is illustrated as a function of extracellular calcium (0.5 mM to 2.0 mM Ca). 4F2 treated cells (squares) prepared from a parathyroid adenoma showed a marked decrease in PTH secretion in low extracellular calcium (0.5 mM to 0.75 mM Ca), which was equivalent to half-maximal hormone secretion by untreated dispersed parathyroid cells (circles; i.e. at 1.25 mM extracellular calcium). At maximally suppressive levels of calcium outside (2.0 mM), PTH release was virtually the same for both treated and untreated cells. Hormone secretion by LC7-2 (triangles) treated cells was indistinguishable from untreated cells. b. Dispersed parathyroid cells were loaded by incubation in 15 µM acetoxymethylester Quin-2, then treated with Mabs 4F2 or LC7-2. Compared to controls, Mab 4F2 treated cells displayed consistantly higher cytosolic calcium levels at all extracellular calcium concentrations examined (e.g. 170% to 300% of controls at 0.5 mM and 2.0 mM Ca, respectively). Cytosolic calcium levels in 4F2 treated cells in low calcium were approximately equal to cytosolic calcium in LC7-2 treated cells or untreated controls in high calcium. The elevated cytosolic calcium values in 4F2 treated parathyroid cells likely account for the observed reduction in PTH secretion.

sensitive dye Quin-2 to measure intracellular free calcium, showed that the inhibition of PTH secretion in 4F2 treated cells was associated with a concomitant increase in cytosolic calcium (Cai) of 188% in 0.5 mM calcium. These values also approached Cai levels in control cells incubated in high calcium (Fig. 10b). Mab controls, P3X63, which did not bind to dispersed parathyroid cells, and Mab LC7-2 (which recognizes a different - light chain - epitope of the same cell membrane glycoprotein as 4F2), did not alter PTH secretion or Cai. These data suggest that Mab 4F2 binding to its cell surface antigen inhibits PTH secretion of human adenomatous parathyroid cells in vitro, and that the alterations in secretory function could be accounted for by attendant increase in Cai. Thus the 4F2/LC7-2 antigen may be an important component of the calcium sensing and/or signal transducing mechanism in cells expressing this glycoprotein. In contrast to parathyroid, preliminary examination of other cell types (monocytes, HSB-2 T lymphocytes, and islet cells) showed that Mab 4F2 (but not LC7-2) binding caused a decrease in cytosolic calcium (57).

Using the same in vitro dispersed parathyroid cell system, in preliminary studies, binding of Mabs HISL-9 and -14 to their corresponding target antigenic epitopes has been found to stimulate, Mab HISL-5 to inhibit PTH secretion. These studies serve to illustrate the usefulness of Mabs as probes to elucidate complex molecular mechanisms involved in normal and pathological endocrine cell function.

MONOCLONAL ISLET CELL ANTIBODIES: APPLICATIONS IN DIAGNOSTIC PATHOLOGY

Application of immunocytochemical techniques to the solution of diagnostic problems in the surgical pathology of neoplasms has progressed greatly in recent years. In contrast to conventional polyclonal antibodies, Mabs represent pure homogenous antibody reagents of defined epitope specificity, which can be produced in unlimited amounts and have therefore the potential as ideal probes for immunocytochemical techniques for diagnostic purposes. Unfortunately, the application of Mabs in routine surgical pathology is limited since many of the epitopes recognized by the Mabs are destroyed by fixation procedures and cannot be detected by immunocytochemistry on formalin fixed and routinely processed specimens.

To investigate their potential as a tool in the immunocytochemical identification of neuroendocrine differentiated tumors, we tested the above listed series of monoclonal islet antibodies for their immunoreactivity on formalin- fixed and paraffin-embedded tissues. With the exception of Mab HISL-19 (Figs. 5, 6), none of the other antibodies reacted with their corresponding antigen on formalin-fixed and paraffin embedded sections of human pancreatic tissues. In subsequent studies the immunoreactivity of Mab HISL-19 was evaluated on a series of 124 neuroendocrine und non-neuroendocrine tumors by an indirect peroxidase technique (48; Tab.4). Mab HISL-19 reacted

Table 4. Immunohistochemical reactivity of Mab HISL - 19 with various neuroendocrine tumors

Tumor type	n	Reactivity
Pituitary adenoma	6	+
Pheochromacytoma	2	+
Neuroendocrine carcinoma of the skin	4	+
C cell carcinoma of the thyroid	8	+
Paraganglioma of the carotid body	3	+
Insuloma	10	+
Carcinoid	8	+
Parathyroid adenoma	3	–
Melanoma	4	+/–
Neuroblastoma	3	+/–
Oat cell carcinoma of the lung	2	+/–

strongly with all insulomas (Fig. 11a), carcinoids (Fig. 11b), pituitary adenomas, paragangliomas, pheochromocytomas, thyroid C cell carcinomas (Fig. 12a) and neuroendocrine carcinomas of the skin tested (Fig. 13c). The latter represent a group of

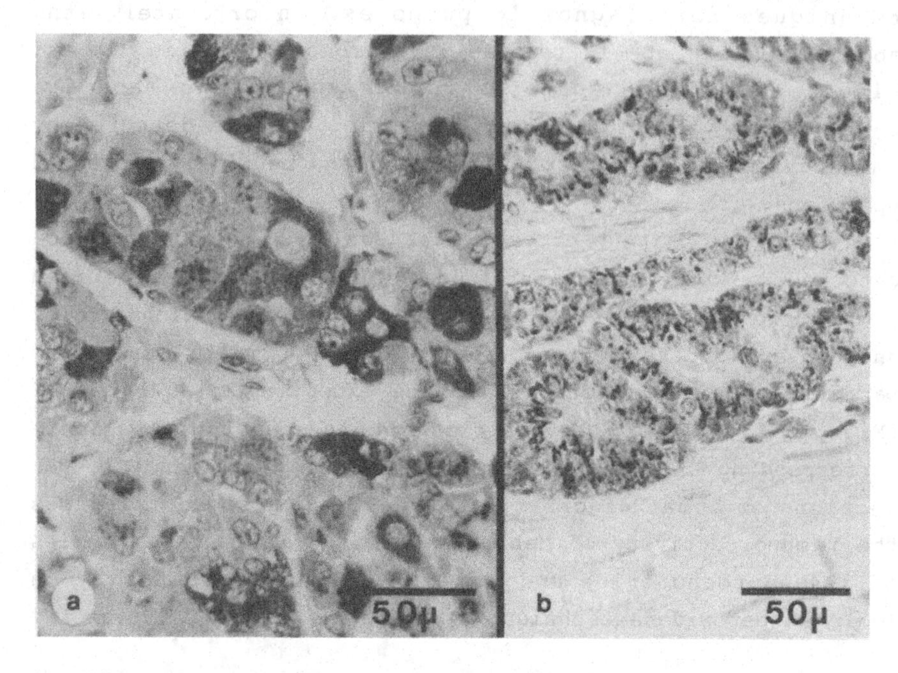

Fig. 11. Immunoperoxidase staining of a human insuloma (a) and a carcinoid tumor (b) with Mab HISL-19 (paraffin sections, counterstained with hematoxilin).

skin neoplasms (Merkel cell tumors) with features of neuroendocrine differentiation and are believed to arise from the Merkel cell (58, 59), which also expresses the HISL-19 antigen (60; Fig. 13b). The well known variability in the light microscopic appearance of these tumors may lead to considerable difficulties in distinguishing them from other skin neoplasms such as malignant lymphomas, melanomas and even poorly differentiated squamous cell carcinomas. The immunocytochemical demonstration of neuroendocrine differentiation antigens such as the HISL-19 antigen is therefore essential in the diagnosis of Merkel cell tumors (60). In contrast, melanomas, bronchial oat cell carcinomas and neuroblastomas, which are known to have neuroendocrine features as evidenced by their content of NSE and the presence of neurosecretory granules in the cytoplasm, were found to show only weak reactivity in a few tumor cells or be unreactive with Mab HISL-19. These findings may reflect a loss of antigen expression or alteration of the antigenic determinant detected by Mab HISL-19 during the neoplastic process, since the cells from which these tumors are believed to originate are all reactive with Mab HISL-19 (i.e. melanocytes, neuroendocrine cells of the bronchial mucosa, adrenal medulla). Parathyroid adenomas (as well as normal parathyroid chief cells) did not express the antigen detected by Mab HISL-19. It seems noteworthy that the inclusion of the parathyroid into the diffuse neuroendocrine system is still controversial, since parathyroid chief cells lack the obligatory "amine precursor uptake and decarboxylation" characteristics (61).

Table 5. Immunohistochemical reactivity of monoclonal antibody HISL-19 with various nonneuroendocrine tumors (positive reactivity is restricted to a supranuclear dotlike staining pattern)

Tumor type	n	Reactivity
Adenocarcinoma of the colon	8	–
Adenocarcinoma of the stomach	5	–
	1	+
Adenocarcinoma of the pancreas	4	–
Acinus cell carcinoma of the pancreas	1	–
Adenocarcinoma of the endometrium	3	–
	1	+
Adenocarcinoma of the bronchus	4	–
Ductal carcinoma of the breast	6	–
Lobular carcinoma of the breast	2	–
	1	+
Squamous carcinoma of the skin	4	–
Transitional carcinoma of the bladder	4	–
Papillary and follicular carcinoma of the thyroid	8	–
Renal adenocarcinoma	3	–
Adrenal cortical adenoma	2	–
Malignant lymphoma	5	–
Leiomyosarcoma	3	–
Leiomyoma of the uterus	3	–
Fibroma	3	–

In order to assess the specificity of the immunoreactivity of Mab HISL-19 to neuroendocrine cells and their related neoplasms, we tested various non-neuroendocrine neoplasms for their reactivity with Mab HISL-19. With a few exceptions, Mab HISL-19 did not react with tumors of this group including adenocarcinomas from various sites, squamous cell carcinomas of the skin, transitional carcinomas of the urinary bladder, thyroid carcinomas of follicular cell origin, adrenal cortical adenomas, malignant lymphomas and various benign and malignant soft tissue tumors (Table 5). Positive

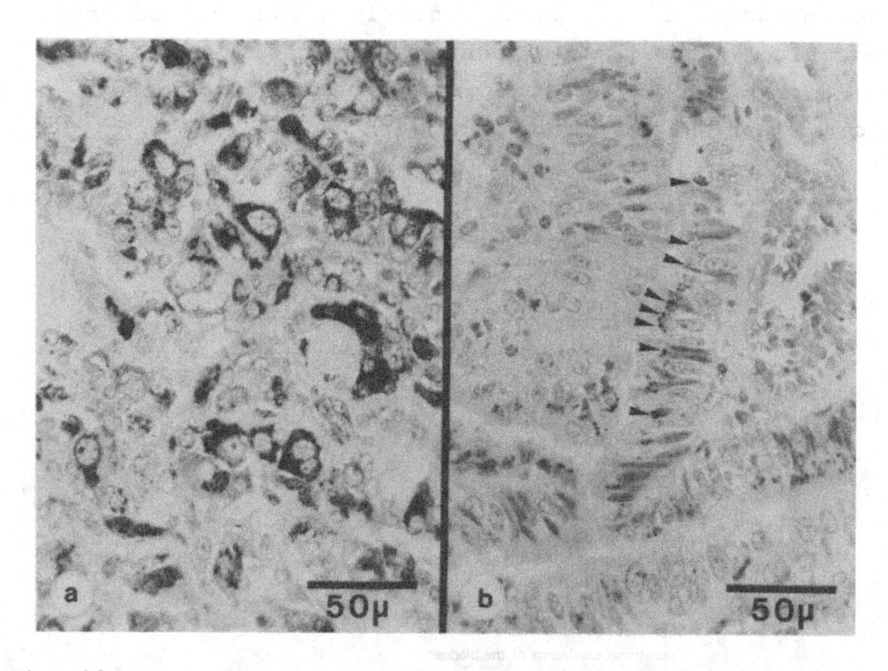

Fig. 12. a. Immunoperoxidase staining of a medullary (C cell) carcinoma of the thyroid with Mab HISL-19. b. Mab HISL-19 immunoreactivity in a human gastric adenocarcinoma revealing supranuclear dot like structures (arrowheads) in some tumor cells (paraffin sections, counterstained with hematoxilin).

staining by Mab HISL-19 was observed only in one adenocarcinoma of the stomach (Fig. 12b), one adenocarcinoma of the endometrium and one breast carcinoma (infiltrating lobular type). The cytologic reactivity pattern was different from the strong granular staining of neuroendocrine cells and was restricted to a small intracytoplasmic paranuclear dot like structure. A similar staining pattern was observed in some normal tissues such as small ducts of salivary glands, the

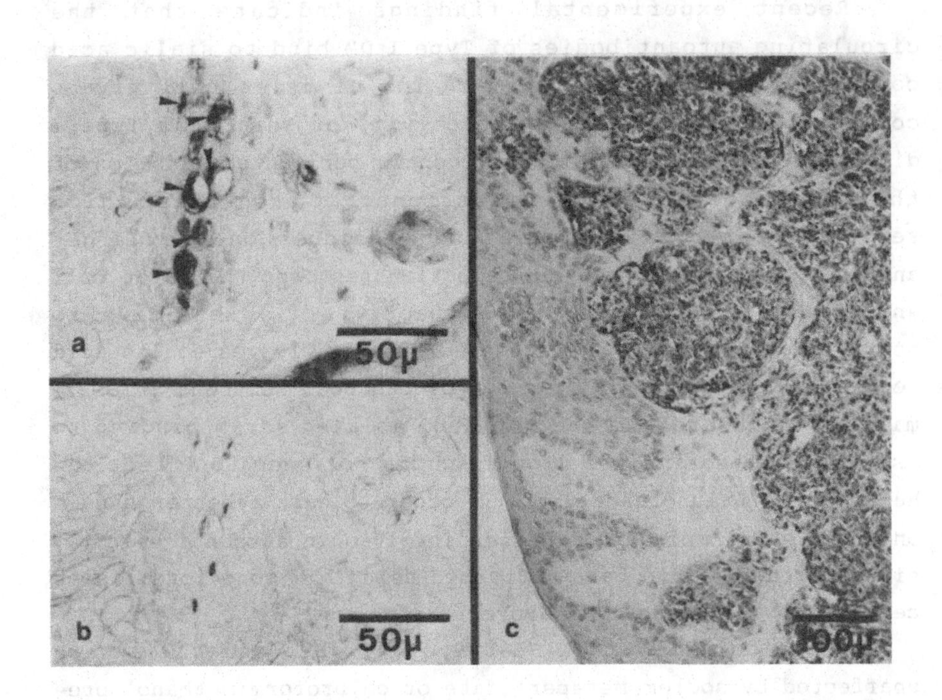

Fig. 13. a. and b. Normal bovine skin stained by an immunoperoxidase technique. The section was reacted first for the antigen detected by Mab HISL-19 (b) yielding a dot-like immunoreactive product in some cells lying above the basement membrane of the epidermis (indirect peroxidase technique). After elution of the antibodies with glycine-HCl buffer, the same section was reacted for NSE (PAP technique) identifying the Mab HISL-19 reactive cells as Merkel cells. The characteristic ring-like appearance of the NSE immunoreaction is in contrast to the still visible Mab HISL-19 binding sites (arrowheads; paraffin sections, no counterstain). c. Neuroendocrine (Merkel cell) carcinoma of the skin immunostained with Mab HISL-19 (paraffin section, counterstained with hematoxilin).

respiratory epithelium and the gallbladder mucosa. These findings suggest that Mab HISL-19 finds potential applications in diagnostic pathology as an indicator for neuroendocrine neoplasms.

MONOCLONAL ISLET CELL ANTIBODIES: RELEVANCE TO TYPE I DIABETES MELLITUS AND ORGAN SPECIFIC AUTOIMMUNITY?

Recent experimental findings indicate that the circulating autoantibodies of Type I DM bind to sialic acid containing carbohydrate side-chains of islet cell glycoconjugates (62). Islet cell reactivity of sera from Type I diabetic subjects is ablated by gentle periodate oxidation of the substrate pancreatic sections, a procedure which selectively modifies sialic acid residues on the target antigens. Sodium borohydrate reduction restores the islet cell antigenicity. Moreover, this reactivity is sensitive to neuraminidase treatment which specifically cleaves sialic acid residues. Furthermore, immersion of pancreas sections in a 2:1 mixture of chloroform and methanol ablated serum binding to islets. Limited pronase digestion did not seem to affect the human serum immunoreactivity. In order to gain a better handle on the precise molecular species involved in autoantigenicity, similar studies have been performed using the monoclonal islet cell antibodies described above.

The binding of Mabs HISL-9, -14, 4F2, and LC7-2 were unaffected by sodium metaperiodate or chloroform:methanol pretreatment of the antigenic pancreatic substrate (63). The binding of these autoantibodies was, however, diminished by limited pronase digestion of pancreas sections, indicating that the epitope for each of these antibodies is protein born. The binding of Mab HISL-19 was ablated by both periodate oxidation and chloroform:methanol treatment of pancreas sections, and is also greatly diminished by pronase digestion; these are properties consistent with previous studies identifying HISL-19 antigen(s) as glycoprotein. The chloroform:methanol sensitivity of the HISL-19 antigen raises an interesting possibility: while some glycoproteins are known to be soluble in chloroform:methanol, it is also possible, that a proportion of HISL-19 reactivity may arise due to similar epitopes born on glycolipids. The binding of Mab 3G5 (which is known to bind to glangliosides antigens) was ablated by both periodate oxidation and chloroform:methanol pretreatment of pancreatic sections. This binding activity was, however, insensitive to pronase digestion. Further biochemical and structural analysis is mandatory to establish the relationship between the various Mab defined islet cell antigens and the bona fide autoantigens actually involved in organ specific autoimmunity.

REFERENCES

1. Cahill, G.F., McDevitt, H.O. New Engl. J. Med. 299: 1439-1465, 1978.
2. Srikanta, S., Ganda, O.P., Eisenbarth, G.S., Soeldner, J.S. New Engl. J. Med. 308: 322-325, 1983.
3. Srikanta, S., Ganda, O.P., Jackson, R.A., Gleason, R.E., Kaldany, A., Garovoy, M.R., Milford, E.L., Carpenter, C.B., Soeldner, J.S., Eisenbarth, G.S. Ann. Int. Med. 99: 320-326, 1983.
4. Ganda, O.P., Srikanta, S., Brink, S.J., Morris, M.A., Gleason, R.E., Soeldner, J.S., Eisenbarth, G.S. Diabetes 33: 516-521, 1984.
5. Srikanta, S., Ganda, O.P., Gleason, R.E., Jackson, R.A., Soeldner, J.S., Eisenbarth, G.S. Diabetes 33: 717-720, 1984.
6. Irvine, W.J., Gray, R.S., Steel, J.M., In: Immunology of Diabetes (Ed. W. J. Irvine), Edinburgh Teviot Scientific Publications Ltd., 1980, pp. 117-154.
7. Gorsuch, A.N., Spencer, K.N., Lister, J., McNalli, J.M., Dean, B.M., Bottazzo, G.F., Cudworth, A.G. Lancet 2: 1363-1365, 1981.
8. Betterle, C., Zanette, F., Trengo, A., Trevisan, A. Lancet 1: 284-285, 1982.
9. Cahill, G.F., McDevitt, H.O. New Engl. J. Med. 304: 1454-1465, 1981.
10. Eisenbarth, G.S. In: Textbook of Endocrinology (Eds. DeGroot L.J.), Grune & Stratton, New York, 1985, in press.
11. Jackson, R., Rassi, N., Crump, T., Haynes, B., Eisenbarth, G.S. Diabetes 30: 887-889, 1981.
12. Jackson, R., Bowring, M., Morris, M., Haynes, B., Eisenbarth, G.S. 63rd Annual Meeting of Endocrine Society, Cincinnati, OH. 450a, 1981.
13. Naji, A., Silvers, W.K., Bellgrau, D., Barker, C.F. Science 213: 1390-1392, 1981.
14. Bottazzo, G.F., Florn-Christensen, A., Doniach, D. Lancet 2: 1279-1283, 1974.

15. Lernmark, A., Freedman, Z.R., Hofmann, C., Rubenstein, A.H., Steiner, D.F., Jackson, L., Winter, R.J., Traisman, H.S. N. Engl. J. Med. 299: 375-380, 1978.
16. Baekkeskov, S., Nielsen, J.H., Marner, B., Bilde, T., Ludvigsson, J., Lernmark, A. Nature 298: 167-169, 1982.
17. Köhler, G., Milstein, C. Nature (London) 256: 495-497, 1975.
18. Eisenbarth, G.S., Jackson, R.A. Endocrin Rev. 3: 26-39, 1982.
19. Srikanta, S., Eisenbarth, G.S. Lancet 1: 1176-1177, 1984.
20. Srikanta, S., Ganda, O.P., Jackson, R.A., Brink, S.J., Fleischnick, E., Yunis, E., Alper, C., Soeldner, J.S., Eisenbarth, G.S. Diabetologia 27 (suppl): 146-148, 1984.
21. Srikanta, S., Ganda, O.P., Rabizadeh, A., Soeldner, J.S., Eisenbarth, G.S. N. Engl. J. Med. 313: 461-464, 1985.
22. Kanatsuna, T., Baekkeskov, S., Lernmark, A., Ludvigsson, J. Diabetes 35: 520-524, 1982.
23. Eisenbarth, G.S., Oie, H., Gazdar, A., Chick, W.L., Schultz, J.A., Scearce, R.M. Diabetes 30: 226-230, 1981.
24. Eisenbarth, G.S., Linnenbach, A., Jackson, R.A., Scearce, R., Croce, C. Nature 300: 264-267, 1982.
25. Eisenbarth, G.S., Jackson, R.A., Srikanta, S., Powers, A.C., Buse, J.B., Rabizadeh, A., Mori, H. In: Immunology in Diabetes (Eds. D. Andreani, U. DiMario, K.F. Federlin, L.G. Heding), Kimpton Medical Publications, London, 1984, pp. 143-175.
26. Crump, A., Scearce, R., Dobersen, M., Kortz, W.J., Eisenbarth, G.S. J. Clin. Invest. 70: 659-666, 1982.
27. Powers, A.C., Rabizadeh, A., Akeson, R., Eisenbarth, G.S. Endocrinology 114: 1-6, 1984.
28. Scearce, R.M., Eisenbarth, G.S. In: Methods in Enzymology (Eds. S.P. Colowick, N.O. Kaplan), Academic Press, New York, 1983, 103, pp. 459-469.
29. Srikanta, S., Ricker, A., Rabizadeh, A., Eisenbarth, G.S. Clin. Res. 32: 409A, 1984.
30. Srikanta, S., Ricker, A., Rabizadeh, A., Eisenbarth, G.S. Excerpta Medica 1825: 1173, 1984.
31. Srikanta, S., Eisenbarth, G.S. In: Methods in Diabetes Research, (Eds. J. Larner, S.L. Pohl), New York, John Wiley & Sons, Inc., vol. I, 1984, pp. 195-208.
32. Srikanta, S., Krisch, K., Eisenbarth, G.S. Diabetes 35: 300-305, 1986.
33. Satoh, J., Prabhakar, B.S., Haspel, M.V., Fellner, F.G., Notkins, A.L. New Engl. J. Med. 309: 217-220, 1983.
34. Satoh, J., Essani, K., McClintock, P.R., Notkins, A.L. J. Clin. Invest. 74: 1526-1531, 1984.
35. Haspel, M.V., Onodera, T., Prabhakar, B.S., Horita, M., Suzuki, H., Notkins, A.L. Science 220: 304-306, 1983.
36. Haspel, M.V., Onodera, T., Prabhakar, B.S., McClintock, P.R., Essani, K., Ray, U.R., Yagihashi, S. Notkins, A.L. Nature 304: 73-76, 1983.
37. Alejandro, R., Sheinvold, F.L., Vaerewyck-Hyjek, R.A., Pierce, M., Paul, R., Mintz, D.H. J. Clin. Invest. 74: 25-38, 1984.

38. Sai, P., Kremer, M., Barriere, P. Hybridoma 3: 131-140, 1984.
39. Yokono, K., Shii, K., Hari, J., Yaso, S., Imanura, Y., Ejiri, K., Ishihara, K., Fujii, S., Kazumi, T., Tanaguchi, H., Baba, S. Diabetologia 26: 379-385, 1984.
40. Brogren, C., Hirsch, F., Wood, P., Druet, P., Poussier, P. Diabetologia 29: 330-333, 1986.
41. Witt, S., Hehmke, B., Dietz, H., Ziegler, B., Hildmann, W., Ziegler, M. Biomed. Biochim. Acta 44: 117-121, 1985.
42. Vissing, H., Papadopulos, G., Lernmark, A. Scand. J. Immunol. 23: 425-433, 1986.
43. Kortz, W.J., Reiman, T.H., Bollinger, R.R., Eisenbarth, G.S. Surg. Forum 33: 354-356, 1982.
44. Eisenbarth, G.S., Shimizu, K., Bowring, M.A., Wells, S.A., Proc. Natl. Acad. Sci. 79: 5066-5070, 1982.
45. Srikanta, S., Telen, M., Reiman, T., Dolinar, R., Krisch, K., Haynes, B.F., Eisenbarth, G.S. manuscript in preparation.
46. Phillips, D.R., Morrison, M. Biophys. Res. Commun. 40: 284, 1982.
47. Polak, J.M., Bloom, S.R. J. Histochem. Cytochem. 27: 1398-1400, 1979.
48. Krisch, K., Buxbaum, P., Horvat, G., Krisch, I., Neuhold, N., Ulrich, W., Srikanta, S. Am. J. Pathol. 123: 100-108, 1986.
49. Schmechel, D., Marangos, P.J., Brightman, M. Nature 276: 834-836, 1978.
50. Blaschko, H., Comline, R.S., Schneider, F.H., Silver, M., Smith, A.D. Nature 215: 48-59, 1967.
51. Eisenbarth, G.S., Haynes, B.F., Schroer, F.A., Fauci, A.S. J. Immunol. 124: 1237-1244, 1980.
52. Srikanta, S., Telen, M., Reiman, T., Dolinar, R., Krisch, K., Haynes, B.F., Eisenbarth, G.S. manuscript in preparation.
53. Eisenbarth, G.S., Walsh, F., Nirenberg, M. Proc. Natl. Acad. Sci. USA 76: 4913-4915, 1979.
54. Rassi, N., Shimuzu, K., Eisenbarth, G.S., Lebovitz, H.E. Clin. Res. 29: 734A, 1981.
55. Hari, J., Yokono, K., Yonezawa, K., Amano, K., Yaso, K., Imamura, Y. Diabetes 35: 517-522, 1986.
56. Posillico, J.T., Srikanta, S., Eisenbarth, G.S., Brown, E.M., Clin. Res. 33(2): 573A, 1985.
57. Posillico, J.T., Srikanta, S., Brown, E.M., Eisenbarth, G.S. Clin. Res. 33(2): 385A, 1985.
58. Tang, C.K., Toker, C. Cancer 42: 2311-2321, 1978.
59. Sibley, R.K., Rosai, J., Foucar, E., Dehner, L.P., Bosl, G., Am. J. Pathol. 4: 211-221, 1980.
60. Buxbaum, P., Horvat, G., Gamper, C., Krisch, K. Cancer Detect. Prevent. 10: 1986, in press.
61. Pearce, A.G.E., Takor, T. Peptides of the brain and gut. Fed. Proc. 1979, pp. 2288-2294.
62. Nayak, R.C., Omar, M.A.K., Rabizadeh, A., Srikanta, S., Eisenbarth, G.S. Diabetes 34: 617-619, 1985.
63. Nayak, R.C., Colman, P.G., Eisenbarth, G.S. Clin. Immunol. Allerg. 7: 1987, in press.

4

A METHOD FOR LONG-TERM MAINTENANCE OF FIBROBLAST-FREE GLUCOSE-SENSITIVE ISLET CELL MONOLAYERS FROM ADULT RATS

N. KAISER[1], P.A.W. EDWARDS[2], AND E. CERASI[1]

[1]Department of Endocrinology and Metabolism, The Hebrew University Hadassah Medical Center, Jerusalem, Israel, and [2]Department of Pathology, University of Cambridge, Cambridge, U.K.

ABSTRACT

Long-term survival of functioning monolayer cultures of adult rat pancreatic islets was obtained by plating islet fragments on dishes coated with extracellular matrix derived from bovine corneal endothelial cells. Fibroblast overgrowth was supressed by the use of a monoclonal antifibroblast antibody or with sodium ethylmercurithiosalicylate. Sensitivity to acute stimulation by glucose was preserved for up to 7 weeks in culture. Further functional competence of the B cells in these monolayer cultures was indicated by their response to various inhibitors and stimulators of insulin secretion. The cultured islets were capable of synthesizing DNA. These findings suggest that the use of ECM-coated plates and removal of the contaminating fibroblasts enhance the long-term survival of adult pancreatic islets, with retained functional capacity. This system seems therefore well suited for studying islet cell physiology and pathology under controlled chronic in vitro conditions.

INTRODUCTION

Insulin-dependent, type I diabetes is caused by a massive reduction of the pancreatic B cell mass due to a chronic destructive process which evolves subclinically during a considerable time (1,2). Intensive research in recent years has led to the concept that in genetically susceptible individuals, the combined effect of viral and/or chemical damage and an autoimmune reaction causes B cell destruction (3). In many cases appearance of islet cell antibodies precedes glucose intolerance by many years (4,5), while some antibody-positive individuals may not become diabetic within a life-time. Since above a given threshold, even a fraction of the normal B cell mass may be sufficient to maintain normal glucose homeostasis (6), it is reasonable to suppose that the rate by which the

Becker, Y (ed), Virus Infections and Diabetes Melitus. ISBN 0-89838-970-4.

islet repairs itself following an autoimmune attack may be of crucial importance for the development of insulin-dependent diabetes.

Potentially, the deterioration of glucose homeostasis in affected individuals could be prevented if the capacity of the B cell to regenerate was increased. For obvious reasons very little information exists regarding B cell replication in man (7). From animal experiments it is known that only a small fraction of the B cells from fetal and neonatal rats enters the cell cycle (8). It is known that B cell replication is influenced by age, metabolic status, various hormones, growth factors, and nutrients (8-13). Most important, it has recently been shown that the proliferative capacity of islets is genetically determined (14).

In order to answer the basic questions related to islet cell replication and its significance for the development of type I diabetes, it is imperative that research is performed under controlled in vitro conditions in islet cell cultures derived from adult donors.

Techniques for culturing pancreatic islets have recently improved considerably. Islets can be cultured free-floating in suspension (15,16). This method allows the isolated islets to maintain their original three dimensional organization and inter-cellular relationships. A similar organization can also be obtained by culturing islet cells embedded in collagen gels which promote the reorganization of cells into islet-like organoids (17). Alternatively, islets can be cultured as monolayers utilizing intact islets or isolated islet cells and a variety of supporting surfaces and treatments to promote islet cell attachment, monolayer formation and long-term survival (18-25). The tissue used for obtaining monolayer culture usually originated from the fetal or neonatal pancreas, with very few investigators being successful in obtaining monolayer cultures from the pancreatic tissue of adult donors. An added complication to the establishment of islet monolayers is that the culture is rapidly overgrown by fibroblasts normally contaminating the islet cell preparations (26,27).

We were recently able to obtain monolayer cultures of adult rat islets by plating isolated islets on plastic plates coated with extracellular matrix (ECM) derived from bovine corneal endothelial cells (28). The ECM was shown to promote cell attachment, proliferation and differentiation of a variety of cells (29). The problem of fibroblast overgrowth was circumvented either by the use of a monoclonal antifibroblast antibody which in the presence of complement is cytotoxic (30), or by treatment with sodium ethylmercurithiosalicylate (thimerosal)(26).

This report describes the technical aspects of the culture method and the initial characterization of the cultured cells.

MATERIALS AND METHODS

Isolation and culture of islets

The islets used for culture are prepared by the collagenase digestion technique adapted from Lacy and Kostianovsky (31).

Male rats from the Hebrew University Sabra strain, 6-8 week old, were used for all experiments. The pancreata from two rats were quickly removed and placed in a Petri dish containing Hank's balanced salt solution (HBSS), pH 7.4. After trimming the tissue free from fat and connective tissue, the pancreata were minced with scissors. The minced tissue was transfered into a conical Pyrex tube and washed 3 times with HBSS. Following the washes 10 mg of collagenase (Serva, Heidelberg, FRG) in 5 ml of HBSS was added and the tube was shaken vigorously in a water bath at 37° for 5-10 min until a homogenous suspension was obtained. After a brief centrifugation, the pellet which contains islets and remaining clumps of exocrine tissue was washed 4 times in 10 ml ice-cold HBSS. Following the last wash, the pellet was resuspended in HBSS at 4° and islets were collected under a stereomicroscope using a micropipette. Only the smallest islets (islet fragments) were collected, as they were shown in preliminary experiments to attach faster to the culture plates and form monolayers earlier. The isolated islets, which still contained some remnants of exocrine tissue and cell debris, were purified by a second isolation under the stereomicroscope and collected into a sterile conical tissue culture tube containing serum-free tissue culture medium. From this step on all procedures were done under strict sterile conditions in a tissue culture hood. The isolated islets were washed 3 times at room temperature by resuspension in RPMI 1640 medium containing antibiotics (100 U/ml penicillin and 100 µg/ml streptomycin) and centrifugation at 150 x g for 5 min. Following the last wash, the islets were resuspended in RPMI 1640 medium containing 10% fetal bovine serum and antibiotics at 37° and plated into 10-15 35 mm tissue culture plates coated with extracellular matrix (ECM). The cultures were maintained at 37° under 10% CO_2 in air, and the culture medium was changed twice weekly.

Preparation of ECM coated plates

To obtain ECM-coated plates, bovine corneal endothelial cells were prepared and cultured in presence of fibroblast growth factor as described by Gospodarowicz et al (29). Following 14 days in culture, the cells were removed by a brief exposure to 0.5% Triton X-100 in 25 mM ammonium hydroxide solution. The ECM-coated plates were washed in phosphate buffered saline (PBS), pH 7.4 and stored filled with PBS for up to 2 months at 4°. Before use for islet culture, the PBS was removed and the plates were washed once with culture medium. In some of our experiments, ECM-coated plates obtained from a commercial source (IBT, Jerusalem, Israel) were used with similar results.

Fibroblast removal

Fibroblasts were removed either by use of antifibroblast monoclonal antibody, or with thimerosal.

Islet cultures were treated routinely before each change of medium as described by Edwards et al (30). After removing the culture medium, the cells were washed in cold serum-free Hepes (10 mM) buffered RPMI 1640, and incubated at 4° for 1 hr with the same medium containing 1:500 dilution of antibody FIB 86.3. After 2 rinses with medium, 1:20 dilution of rabbit serum in Hepes buffered medium was added and the plates were incubated at 37° for 2 hr. Following the incubation, cultures were rinsed twice and regular growth medium was added.

Fibroblast removal with thimerosal was performed as described by Bratten et al (26). On day 3 or 4 after plating, as fibroblasts started to emerge, the cultures were exposed to 1 μg thimerosal per ml of serum containing culture medium for 1-2 days, after which the medium was replaced by the regular growth medium. Although sometimes a single treatment was sufficient to obtain fibroblast-free cultures, the treatment may be repeated at a later time of culture as needed.

Insulin secretion by islet monolayers

The function of the B cells was assessed at different times of culture by measuring glucose-stimulated insulin release and its modulation by various stimulators and inhibitors of secretion. The following experimental protocol was used for all studies. The culture medium was replaced by 1 ml of Krebs-Ringer bicarbonate (KRB) buffer containing 10 mM Hepes, 3.3 mM glucose and 0.25% bovine serum albumin (BSA), pH 7.4. After 60 min of preincubation at 37° under 10% CO_2 atmosphere, the buffer was discarded and replaced by 1 ml of the same buffer. At the end of further 60 min, the buffer was collected and used for determining the basal rate of hormone secretion. This buffer was replaced by KRB-Hepes-BSA buffer containing the test substances, and the incubation continued for another 60 min, at the end of which the buffer was collected for insulin determination. The collected buffers were centrifuged at 4° (15 min, 150 x g) to remove dead cells and cell debris, and the supernatants frozen for hormone assay.

Insulin determination

Insulin was determined by a standard radioimmunoassay technique using guinea pig anti-porcine insulin antibodies (Serono Diagnostics, Coinsins, Switzerland) and 20% PEG to separate bound and free hormone. Rat insulin was used as a standard.

Autoradiography

DNA synthesis in monolayer cultures was determined by incorporation of ^{3}H-thymidine followed by autoradiography. Cultures were incubated for 20 hr with 10 μCi of ^{3}H-thymidine at 37°. The cells were then washed 5 times with ice-cold PBS and fixed with methanol. The plates were coated with Ilford K-5 photographic emulsion, dried and kept for 5 days at room temperature before developing.

RESULTS

The collagenase method used allowed 200-250 islet fragments to be collected from each pancreas. Freshly isolated islet fragments were seen under the phase-contrast microscope to consist of irregular cell clusters of varying size, lacking a well defined capsule. After 2 days in culture on plastic plates (Fig. 1A) the islet fragments remained unattached and their irregular shape was still evident. On the other hand, islet fragments plated on ECM-coated plates were already attached by 3-4 hr, and after 1 day in culture (Fig. 1B) cells started to migrate away from the center of the islet, some fragments being already flattened. By the second day on ECM-coated plates, most of the islets were flattened and the formation of an epitheloid monolayer was evident (Fig. 1C). After 72 hr in culture all islet cells were spread out forming a monolayer (Fig. 1D). At this time some binucleated cells were evident (Fig. 1D & 4). Fibroblast growth usually started to be evident by the 4th day of culture and unless removed, fibroblasts completely surrounded the islet cells, inhibiting further spreading of the cell monolayer.

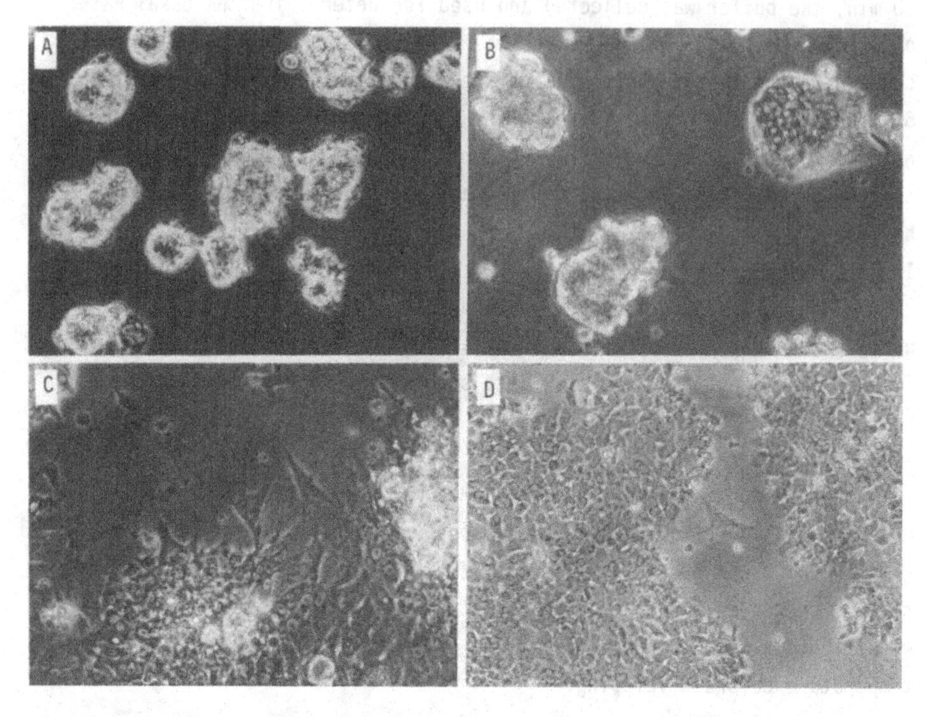

Fig. 1. Fragmented islets cultured on: A) Plastic plate for 48 hr B) ECM-coated plate for 24 hr C) ECM-coated plate for 48 hr D) ECM-coated plate for 72 hr. Phase contrast x 200.

Figure 2A shows an intact islet of Langerhans after 3 days of culture on ECM-coated plate. As opposed to the islet fragments shown in Fig. 1, intact islets did not form a monolayer after 72 hr. Still, these islets were well attached to the plates (Fig. 2A) and started to flatten, with cells migrating away from the center of the islets. Fibroblasts started to appear at the periphery of the intact islets already by the second day of culture. Only after 1 week in culture these large islets formed a circular monolayer with some of the cells in the center of the islet remaining in a multilayer and eventually becoming necrotic (not shown). For this reason we changed our method and collected only the small fragmented islets (Fig. 1). Figure 2B shows the same islet as in Fig. 2A after treatment with antifibroblast monoclonal antibody and complement. While the islet itself remained intact, the fibroblasts were destroyed.

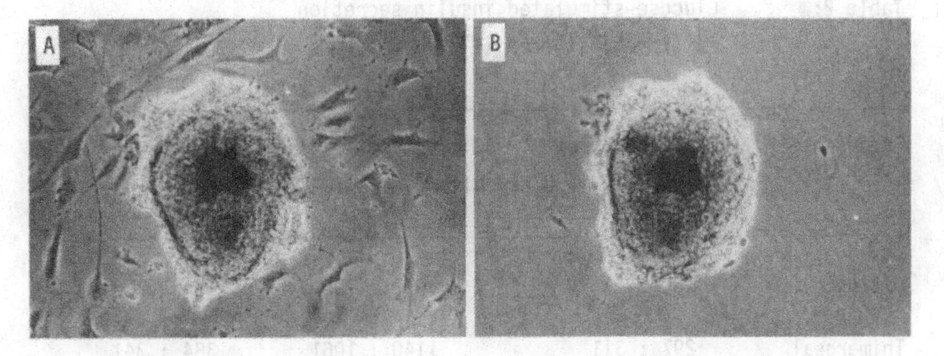

Fig. 2. Intact islet after 3 days in culture on ECM-coated plate. A) Untreated islet surrounded by fibroblasts. B) The same islet after treatment with antifibroblast monoclonal antibody and complement. Phase contrast x 100.

Fibroblast-free cultures obtained by this method and grown on ECM-coated plates maintained a near constant basal (3.3 mM glucose) insulin secretory activity for almost 3 weeks (Table 1) and responded to a glucose stimulus (16.7 mM glucose) with 5.4-fold increase of insulin secretion (Table 2). Unfortunately, due to apparent instability of the hybridomas secreting the antifibroblast antibodies, the cells lost their ability to produce the

Table 1: Insulin secretion in islet monolayers

Days in culture	Secretion rate (μU/hr/culture)
4	180 ± 18
7	181 ± 38
19	149 ; 296

Islet monolayers were grown on ECM-coated plates in RPMI 1640 medium with 10% FCS. Fibroblasts were removed with antifibroblast monoclonal antibody and complement. Mean ± SEM of 4 plates.

Table 2: Glucose-stimulated insulin secretion in fibroblast-free islet monolayers

Treatment	Basal secretion	Glucose stimulation	% change
	(μU/hr/culture)		
Antibody	180 ± 43	972 ± 198	540
Thimerosal	297 ; 311	1140 ; 1061	384 ; 341

One-week old monolayer cultures were either treated with antifibroblast antibody and complement, or exposed to thimerosal (1 μg/ml) from day 2 through 8. Glucose-stimulated insulin secretion was measured by exposure to 3.3 mM glucose for 1 hr followed by 16.7 mM for a second hr. Mean ± SEM of 9 observations.

specific antibody. This loss is thought to be due to non-secreting hybrids, or unrelated hybrids overgrowing the desired hybrids, and may be controlled by recloning the hybrid cells (32). While this procedure was in progress, we tested an alternative way to obtain fibroblast-free islet cultures. The mercurial agent thimerosal was tested for its ability to eliminate fibroblasts associated with adult pancreatic monolayer cultures. Figure 3A shows an 8-day old culture grown on ECM-coated plate. The islet is surrounded by fibroblasts. When the culture was exposed to 1 μg/ml thimerosal from day 2 through 8 the fibroblasts were eliminated without affecting islet monolayer morphology (Fig. 3B). The prolonged exposure to thimerosal did not affect the ability of the islet culture to respond to a glucose stimulus (Table 2); therefore, it was used routinely to eliminate fibroblast contamination. Later it was realized that the time of exposure to thimerosal can be shortened considerably, and at present a 1 to 2 day exposure to 1 μg/ml thimerosal is used routinely to circumvent the problem of fibroblast overgrowth.

The capacity of the B cells to respond to different concentrations of glucose was measured in 1-2 week old cultures following an acute (1 hr) exposure to glucose. Insulin secretion was increased 5 fold when

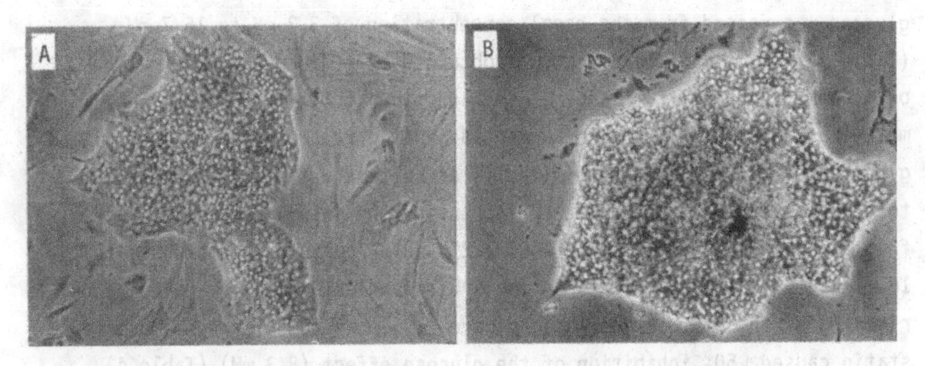

Fig. 3. Islet fragments cultured for 8 days on ECM-coated plates. A) Untreated islet surrounded by fibroblasts. B) Islet exposed to 1 µg/ml thimerosal from day 2 through 8. Phase contrast x 100.

Table 3: Glucose dose-dependency of insulin secretion in the absence and presence of 0.1 mM IBMX

Glucose concentration (mM)	Insulin secretion (% of basal)	
	-IBMX	+IBMX
3.3	100 ± 5	344 ± 20
5.5	242 ± 38	767 ± 102
6.9	306 ± 51	1028 ± 346
8.3	408 ± 82	991 ± 106
16.7	499 ± 68	1367 ± 153

Cultures were 1-2 week old. Plates were exposed for 1 hr to a basal glucose concentration (3.3 mM), followed by a second hr of exposure to increasing concentrations of glucose without IBMX and a third hr with 0.1 mM IBMX. Results are mean ± SEM of 3 observations.

glucose was raised from the basal concentration of 3.3 mM to 16.7 mM (Table 3). Half maximal effect was obtained at a glucose concentration of 5.5 mM. As seen from Table 3, the addition of 0.1 mM 3-isobutyl-1-methylxanthine (IBMX) further augmented insulin release by 3 fold at all glucose concentrations tested. The islet culture was still responsive to glucose at 7 weeks of culture, increasing the insulin secretion rate from 330 ± 27 μU/hr/culture to 1386 ± 297 μU/hr/culture when exposed to 16.7 mM glucose. The effects of the inhibitors somatostatin and epinephrine and of the stimulator glucagon were also studied. Ten μg/ml somatostatin caused ∿50% inhibition of the glucose effect (8.3 mM) (Table 4).

Table 4: Effect of hormones on glucose-induced insulin secretion

Hormone concentration	Basal secretion (3.3 mM glucose)	Stimulated secretion (8.3 mM glucose)	8.3 mM glucose + hormone
Somatostatin			
1 ng/ml	175 ± 24	712 ± 79	876 ± 142
100 ng/ml	200 ± 54	676 ± 90	527 ± 41
10,000 ng/ml	134 ± 67	560 ± 72	280 ± 78
Epinephrine			
1 ng/ml	534 ± 44	1191 ± 171	989 ± 36
100 ng/ml	228 ± 114	1003 ± 119	150 ± 20
10,000 ng/ml	124 ± 47	552 ± 87	83 ± 11
Glucagon			
1 ng/ml	384 ± 155	1486 ± 323	1293 ± 224
100 ng/ml	164 ± 70	704 ± 61	1246 ± 232
10,000 ng/ml	209 ± 137	746 ± 132	1313 ± 90

Islet monolayers were 1-2 week old. One hr exposure to 3.3 mM glucose is taken as the basal secretion rate. The stimulated secretion was measured by a second one hr exposure to 8.3 mM glucose, and the effect of the various hormones by a third hr of exposure to different concentrations of hormones in the presence of 8.3 mM glucose. Results are expressed in μU/hr/culture and represent mean ± SEM of 3-4 plates.

Epinephrine had a maximal inhibitory effect at 100 ng/ml under our experimental conditions. The adult rat cultured islets were also responsive to the stimulatory effect of glucagon, 100 ng/ml causing a further 2-fold increase in insulin secretion as compared to 8.3 mM glucose alone.

Preliminary experiments using 3-day old cultures of adult rat islets suggest that under in vitro conditions the monolayers grown on ECM-coated plates synthesize DNA (Fig. 4). While no quantitative determination of the rate of DNA synthesis could be obtained using our experimental protocol, it appears that 2-3% of the nuclei are labeled by 20 hr exposure to ^{3}H-thymidine.

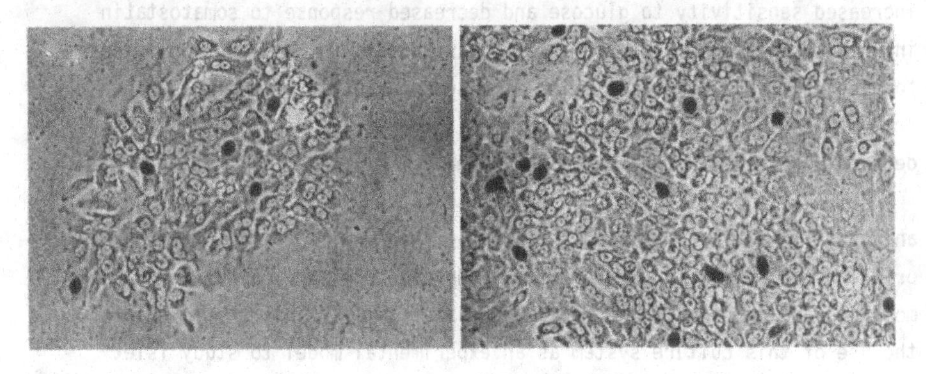

Fig. 4. Autoradiograms of two islet monolayers grown for 3 days on ECM-coated plates and labeled with ^{3}H-thymidine for 20 hr. Phase contrast x 200.

DISCUSSION

The use of monolayer cultures derived from the adult rat pancreas has the advantage over the previously utilized monolayers of neonatal or fetal pancreatic tissue, in that it represents more closely the physiological state of islets in individuals who are susceptible to type I diabetes.

In vivo all epithelial cells capable of proliferation or long-term survival are found in apposition to a layer of mesenchimal cells, separated by a basement membrane. It is clear from this report that the use of ECM secreted by corneal endothelial cells indeed promoted cell attachment and formation of monolayer from fragmented islets of adult rats. The cultured islets were functional for up to 7 weeks in culture as indicated by their ability to respond to an acute glucose stimulus with a 4-fold increase in insulin secretion. The cultures appear to maintain a functional state also as judged by their response to a number of mediators. While the glucose-stimulated insulin secretion showed a higher sensitivity to glucose (ED_{50} of 5.5 mM, Table 3) as compared to freshly isolated adult rat islets (33), or cultured islets from neonates (34), IBMX further augmented the response. The inhibition of the glucose-mediated insulin secretion by epinephrine is consistent with the observation of Marliss et al (35) in monolayers of neonatal rat islets and of Rabinovitch et al in isolated adult rat islets (36). Somatostatin, on the other hand, seems to be less effective in the monolayer cultures as compared to freshly isolated islets (37). Work is in progress to investigate the reasons for the increased sensitivity to glucose and decreased response to somatostatin in this system. The cultured islets maintained also their responsiveness to glucagon. DNA synthesis which is also apparent in vitro in islets from neonatal and fetal rodents (8), still continues in the monolayers derived from the adult pancreatic tissue.

The data obtained so far suggest that the use of ECM-coated plates and the elimination of fibroblasts with either antifibroblast antibodies or thimerosal, promote the formation and long-term survival of functioning monolayer cultures of adult pancreatic islet cells. This should enable the use of this culture system as an experimental model to study islet damage and repair mechanisms under conditions which simulate type I diabetes.

ACKNOWLEDGMENTS

We are grateful to Ariella Tur-Sinai and Yaffa Ariav for their expert technical help. This work was supported by grants from The Juvenile Diabetes Foundation International, and Yad Hanadiv Foundation.

REFERENCES

1. Gepts, W. and DeMey J. Diabetes 27 (Suppl. 1): 251-261, 1978.
2. Gepts, W. and LeCompte, P.M. Am. J. Med. 70: 105-115, 1981.
3. Andreani, D. In: Immunology in Diabetes (Eds. D. Andreani, U. Di Mario, K.F. Federlin and L.G. Heding), Kimpton Medical Publications, London, 1984, pp. 1-20.
4. Gorsuch, A.N., Spencer, K.M., Lister, J., McNally, J.M., Dean, B.M., Bottazzo, G.F. and Cudworth, A.G. Lancet II: 1363-1365, 1981.
5. Srikanta, S., Ganda, O.P., Eisenbarth, G.S. and Soeldner, J.S. N. Engl. J. Med. 308: 322-325, 1983.
6. Hegre, O.D. In: The Diabetic Pancreas (Eds. B.W. Volk and E.R. Arquilla) Plenum Medical Book Co., New York, 1985, pp. 513-542.

7. Campbell, I.L., Harrison, L.C. Christopher, L.J., Colman, P.G. and Ellis, D.W. Am. J. Clin. Pathol. 84: 534-541, 1985.
8. Hellerström C. Diabetologia 26: 393-400, 1984.
9. Nielsen, J.H. Endocrinology 110: 600-606, 1982.
10. Andersson, A. and Sandler, S. In: Immunology in Diabetes (Eds. D. Adreani, U. Di Mario, K.F. Federlin and L.G. Heding), Kimpton Medical Publications, London, 1984, pp. 177-186.
11. Rabinovitch, A., Quigley, C., Russell, T., Patel, Y. and Mintz, D.H. Diabetes 31: 160-164, 1982.
12. Logothetopoulos, J., Valiquette, N. and Cvet, D. Diabetes 32: 1172-1176, 1983.
13. Swenne, I. Diabetes 32: 14-19, 1983.
14. Swenne, I. and Andersson A. Diabetologia 27: 464-467, 1984.
15. Andersson, A. Diabetologia 14: 397-404, 1978.
16. Nielsen, J.H., Brunstedt, J., Andersson, A. and Frimodt-Møller, C. Diabetologia 16: 97-100, 1979.
17. Montesano, R., Mouron, P., Amherdt, M. and Orci, L. J. Cell Biol. 97: 935-939, 1983.
18. Lambert, A.E., Blondel, B., Kanazawa, Y., Orci, L. and Renold, A.E. Endocrinology 90: 239-248, 1972.
19. Ohgawara, H., Carroll, R., Hofmann, C., Takahashi, C., Kikuchi, M., Labrecque, A., Hirata, Y. and Steiner, D.F. Proc. Natl. Acad. Sci. USA 75: 1897-1900, 1978.
20. Meda, P., Hooghe-Peters, E.L. and Orci, L. Diabetes 29: 497-500, 1980.
21. Rabinovitch, A., Russell, T. and Mintz, D.H. Diabetes 28: 1108-1113, 1979.
22. Chick, W.L., Like, A.A. and Lauris, V. Science 187: 847-849, 1975.
23. Bone, A.J. and Swenne, I. In Vitro 18: 141-148, 1982.
24. Thivolet, C.H., Chatelain, P., Nicoloso, H., Durand, A. and Bertrand, J. Expt. Cell Res. 159: 313-322, 1985.
25. Kaiser, N., Tur-Sinai, A., Edwards, P.A.W. and Cerasi, E. Diabetes Res. Clin. Pract. Suppl. 1: S279, 1985.
26. Bratten, J.T., Lee, M.J., Schenk, A. and Mintz, D.H. Biochem. Biophys. Res. Commun. 61: 476-482, 1974.
27. Yoshida, K., Kagawa, S., Murakoso, K., Nakao, K., Shimizu, S., Wakabayashi, S., Haito, K., Tanaka, N., Iha, B. and Matsuoka, A. Horm. Metab. Res. 16: 120-124, 1984.
28. Gospodarowicz, D., Mescher, A.L. and Birdwall, C.R. Expt. Eye Res. 25: 75-87, 1977.
29. Gospodarowicz, D. and Tauber, J.P. Endocr. Rev. 1: 201-227, 1980.
30. Edwards, P.A.W., Easty, D.M. and Foster, C.S. Cell Biol. Intl. Rep. 4: 917-922, 1980.
31. Lacy, P.E. and Kostianovsky, M. Diabetes 16: 35-37, 1967.
32. Edwards, P.A.W. Biochem. J. 200: 1-10, 1981.
33. Grill, V. and Cerasi, E. J. Biol. Chem. 249: 4196-4201, 1974.
34. Dobersen, M.J., Scharff, J.E. and Notkins, A.L. Endocrinology 106: 1070-1073, 1980.
35. Marliss, E.B., Wollheim, C.B., Blondel, B., Orci, L., Lambert, A.E., Stauffacher, W., Like, A.A. and Renold, A.E. Eur. J. Clin. Invest. 3: 16-26, 1973.
36. Rabinovitch, A., Cerasi, E. and Sharp, G.W. Endocrinology 102: 1733-1740, 1978.
37. Ipp, E., Oron, R. and Cerasi, E. Diabetes Res. Clin. Pract. Suppl. 1: S260, 1985.

7. Campbell, [illegible] L., Harrison, [illegible] C., Christopher, L. [illegible] Johnson, [illegible] and [illegible], J. W., Am. J. Clin. Path. 84: [illegible]-641, 1985.
8. Gellerstedt, [illegible] C., Diabetologia 28: [illegible], 1985.
9. Skelton, [illegible] Endocrinology 111: 600-608, 1982.
10. Aggarwal, A., and Sandler, S. In: Immunology in Diabetes (Eds. D. Andreani, U. Di Mario, K. F. Federlin and L. G. Heding). Kimpton Medical Publications, London, 1984, pp. [illegible]-286.
11. Rabinovitch, A., Quigley, C., Russell, T., Patel, Y. and Mintz, [illegible] D. H., Diabetes 31: 160-164, 1982.
12. [illegible] J., Vallejo, [illegible] and Ivey, [illegible] 1152-1158, 1985.
13. Welsh, [illegible] Diabetes [illegible] 144-[illegible]
14. Werne, [illegible] and Andersson, A., Diabetologia 27: [illegible]-[illegible], 1984.
15. Andersson, A., Diabetologia 1: 397-404, 1979.
16. Nielsen, [illegible] Brunstedt, [illegible] Andersson, A. and Frimodt-[illegible] Diabetologia 16: [illegible]-100, 1979.
17. [illegible] J. Cell Biol. [illegible]: 53, 1982.
18. [illegible], A., Blondel, B., Yamazawa, [illegible] Endocrinology [illegible]-[illegible], 1972.
19. Ohgawara, H., [illegible] Hofmann, C., Kumar, [illegible] Kaminski, [illegible] Sci. [illegible] 1978.
20. [illegible] Peters, [illegible] and [illegible] Diabetes [illegible] 19[illegible]
21. Rabinovitch, [illegible] Russell, T. and Mintz, [illegible] Diabetes 28: 1108, 1979.
22. [illegible] and Lacy, [illegible] 8: 809, 19[illegible]
23. [illegible] and [illegible] In Vitro 18: [illegible], 198[illegible]
24. [illegible] Nicholas, [illegible] and Gerhard, [illegible] Exp. Cell Res. 159: 510-[illegible], 1985.
25. Carlsson, [illegible] A., Edwards, R. A. W. and Gerasi, [illegible] Diabetes Res. Clin. Pract. Suppl. 1: [illegible]-275, 1985.
26. [illegible] Lee, M., Clark, A. and Mintz, [illegible] Biophys. Res. Commun. 61: [illegible]-452, 1974.
27. Yoshida, K., [illegible] Murakami, [illegible] Nakabayashi, S., [illegible] Tanaka, [illegible] and Matsuoka [illegible] Hor. Metab. Res. 16: 120-[illegible], 1984.
28. [illegible] Fischer, [illegible] and Bidwell, [illegible] Exp. Eye Res. [illegible] 1977.
29. Gospodarowicz, [illegible] and Tauber, J. P., Endocr. Rev. 1: 201-227, 1980.
30. Edwards, R. A. W., Lesley, D. M. and Foster, [illegible] Cell Biol. [illegible] Vet. [illegible] 21-[illegible], 1985.
31. [illegible] P. E. and Rostamkhani, [illegible] Diabetes [illegible] 1987.
32. Edwards, R. A. W., Nature [illegible] 1977.
33. Griffin, [illegible] and Gerasi, E., J. Biol. Chem. 249: 4198-4201, 1974.
34. Pedersen, [illegible] Schmidt, [illegible] and [illegible] Endocrinology [illegible]: 1070-1073, 1980.
35. [illegible] Blondel, B., [illegible] A. E., Stauffacher, [illegible] and Renold, A. E., Eur. J. Clin. Invest. [illegible]: 18-26, 1973.
36. Rabinovitch, A., Gerasi, E. and Sharp, [illegible] Endocrinology 102: 1118-1130, 1978.
37. Ing, [illegible] Groop, [illegible] and Gerasi, E., Diabetes Res. Clin. Pract. Suppl. 1: S285, 1985.

Clinical considerations of diabetes mellitus

5

CLINICAL ASPECTS OF DIABETES TYPE I AND TYPE II

H. BAR-ON
Diabetes Unit, Hadassah Medical Center, Jerusalem, Israel

ABSTRACT

Diabetes mellitus is a syndrome of various genetic and environmental etiologies. The hyperglycemia and its associated metabolic, functional and pathological sequelae apparently have different mechanisms, depending on the several entities encountered. Two major clinical presentations of diabetes are reviewed in this chapter and their genetic and pathogenetic characteristics are described. Patients with diabetes will eventually develop micro- and macro-vascular complications, making their lives increasingly uncomfortable and distressing. A short report on these grave afflictions, their pathogenesis and course of disease has been given. The relationship between diabetic control and complications is also discussed. Although a lot of progress in understanding diabetes and its therapy has been made in the last six decades, a great deal is still left for future clinical and basic research to accomplish.

INTRODUCTION

The classic symptoms and signs of diabetes mellitus (DM) (1-6) which include polyuria, polydipsia, glycosuria, hyperglycemia, ketosis, loss of weight and coma, were always considered to be due merely to insulin deficiency. Only recently has it been recognized that the syndrome of DM is actually comprised of a heterogeneous group of diseases with widely diverse etiologies, epidemiological presentations, and genetic

Becker, Y (ed), Virus Infections and Diabetes Melitus. Publishing, Boston. ISBN 0-89838-970-4.

predisposition. Two main clinical aspects are encountered in DM, namely the early metabolic derangements and the late pathological changes. It is still debatable whether these basic phases of the disease are causally related. It will be better understood if we consider DM as a multifactorial disorder culminating in insulin inefficiency that gives rise to hyperglycemia - the hallmark of all the various diabetogenic insults. The hyperglycemia and the associated metabolic changes induced by the loss of insulin activity are not only the biochemical features of the disease but might become operative in the development of the pathological complications.

CLASSIFICATION AND DIAGNOSIS

There are two major categories in the new classification (1) of DM namely, IDDM (Type I, insulin-dependent diabetes mellitus, formerly known as juvenile onset DM) and NIDDM (Type II, non-insulin-dependent diabetes mellitus, formerly known as adult onset DM). In addition, other associated conditions have been listed in the new classification that include impaired glucose tolerance (IGT), gestational diabetes mellitus (GDM), and other types of diabetes.

The criteria for establishing DM have been prepared and reviewed, in order to standardize the diagnostic methodology and to achieve international agreement thereon. Currently, the World Health Organization (WHO) and National Diabetes Data Group (NDDG) recommendations are adopted world-wide (see Table 1).

It is apparent that OGTT is unnecessary in a symptomatic patient with even a single random plasma glucose of 11.1 mmol/L (200 mg/dL), or higher. It should be emphasized that when ordering OGTT or performing blood test for glucose levels, one has to rule out many non-diabetic causes

Table 1. Diagnostic criteria for diagnosis of DM and IGT.

	Fasting	1/2, 1 or 1 1/2 hr	2 hr
DM			
WHO	>8.0 (140)	Not required	>11.0 (200)
NDDG	>140 (7.8)	>200 (11.1)	>200 (11.1)
IGT			
WHO	<8.0 (140)	Not required	8-11 (140-200)
NDDG	<140 (7.8)	>200 (11.1)	140-200 (7.8-11.1)

Note: The above are diagnostic criteria for diagnosis of DM and IGT following 75g OGTT (oral glucose tolerance test). Values are for venous plasma in mmol/L or mg/dL.

of impaired glucose tolerance, such as liver disease, kidney disease, other endocrine diseases (Cushing's disease, active acromegaly, thyrotoxicosis, pheochromocytoma) or ingestion of certain drugs (thiazide diuretics, corticosteroids, oral contraceptives, dilantin, alcohol, etc.).

The diagnosis of DM can be confirmed by other indices, as for example the measurement of glycosylated compounds in the blood. Plasma glucose reacts non-enzymatically with the hemoglobin in the red blood cell to form several glycosylated hemoglobins (GHb) or could be attached to plasma proteins. Total concentrations of the glycosylated proteins are correlated with the levels of circulating plasma glucose. Studies to assess the sensitivity of measuring GHb as a predictor of glucose intolerance were mostly disappointing, especially in mild glucose intolerance. However, in moderate to severe intolerance, the estimation of GHb was found useful in diagnosing diabetes.

By analogy with several other endocrinological afflictions, measures of plasma insulin levels would be of diagnostic importance; however, as will be discussed later, glucose intolerance and diabetes might be accompanied by hypo-normo or hyperinsulinemia. At the present time, the determination of the insulin response to glucose load alone has a restricted value in establishing the diagnosis of DM.

Still another way to verify the existence of DM is to perform muscle or kidney biopsy and to study the capillary basement membrane (2). Evidence has accumulated to support the view that basement membrane thickening (BMT) is associated with diabetes. Some authors (3) have claimed that BMT preceded the clinical syndrome, leading them to the conclusion that BMT may be part of the disease complex rather than a result of metabolic control.

INSULIN-DEPENDENT DIABETES MELLITUS (IDDM)

Clinical presentation

The majority of patients presenting with this type of disease are children below the age of 15, with peak ages of onset at 8, 10, 12, and 14. Most of the younger patients are admitted to the hospital shortly after the development of first symptoms. The mean interval between onset of symptoms and establishment of the diagnosis is one month.

Approximately 20% of the children have a history of recent infection, and 22% of the patients enter the hospital in ketoacidosis and/or coma. There is a trend of peak incidence of IDDM in the autumn and winter.

Genetics of IDDM

Recently, it has been suggested that IDDM is associated with certain genes within the histocompatibility complex (HLA) (4). More than 90% of the patients with IDDM express the histocompatibility alleles - DR3 or DR4.

or both. On the other hand, there is a negative association between DR2 and IDDM.

Family studies have confirmed that between identical twins, there is a 50% chance of concordance for the disease, with a 20% chance of developing the illness if two haplotypes are shared by the affected sibling and the unaffected one. If one haplotype is shared, there is a 5% chance, but if neither haplotype is shared, there is only a 1% chance.

Evidence is accumulating to support the idea that there are also non-HLA linked genes predisposing to IDDM. It has been shown that the insulin gene on the short arm of chromosome 11 is flanked by a polymorphic region with three major alleles. In a study in which IDDM patients were compared to non-diabetic subjects, one of the alleles (class 1) was preponderant in diabetic patients. (5). A linkage between IDDM and immunoglobulin gene allotype on chromosome 2 was also reported. Given these data, and especially the concordance of only 50% in identical twins, it is clear that environmental agents are also responsible for the development of IDDM.

Pathology

In chronic IDDM patients, the most prominent finding, clinically and biochemically, is absolute deficiency of insulin. The pancreas of a patient dying of a prolonged insulin deficiency state would show atrophy of the islets of Langerhans with fibrosis and hyalinization with occasional deposition of amyloid-like material. No extractable insulin can be found at this stage. In contrast, in short-term disease, the most chracteristic lesion in IDDM is insulitis (6), with mononuclear cells of the lymphocyte-plasma cell line infiltrating the islets. These infiltrates are either located in the periphery of the islets, or they may actually obscure the islet structure completely.

This inflammatory process around or within the islets is accompanied by beta-cell destruction or degranulation. In some cases of short duration, hypertrophic islets with hydropic or ballooning degeneration of the beta cells might be encountered. Depending on the severity of the disease and the time of the examination of the pancreas, the proportion of islets showing degenerative, inflammatory, atrophic or fibrotic changes varies markedly.

Etiology and pathogenesis

Evidence is accumulating to support the hypothesis that type I DM, like some other endocrine diseases (e.g. myxedema, Addison's disease) results from a continuous and progressive destruction of the insulin-secreting cells by an autoimmune process (7, 8). The immune response is mediated by both humoral and cellular mechanisms.

An enormous amount of recently accumulated clinical and experimental data indicates that, at the clinical onset of the disease, antibodies towards specific antigens of the islet cell domain circulate in the plasma and can be identified. These antibodies exert cytotoxic effects on beta cells and are shown *in vitro* to react with various components of the hormone-producing cells. The mechanisms of the "insulitis" described above (see under Pathology) should involve cell-mediated immune response, and several tests employed in the laboratory confirm this hypothesis. Moreover, peripheral lymphocytes isolated from patients with IDDM produced cytotoxic effects against beta cells obtained from animals.

The existence of an array of islet cell antibodies could be established in the plasma of patients long before the full-blown clinical picture ensued, or could be detected in the plasma of siblings of diabetic patients who had no clinical disease (few of these inflicted sibs will develop diabetes later in the course of their lives). In addition, in some diabetic patients, autoimmune responses directed towards other endocrine organs might be found. It is assumed that multiple autoimmune reactions encountered in families with diabetes are responsible for the "polyendocrine deficiency syndrome" in which several hormone-producing organs fail.

The autoimmune response in IDDM is genetically determined, and more than 95% of type I white diabetics carry the HLA alleles DR3 or DR4, or both. However, one must implicate environmental factors as playing a triggering role in the initiation of the autoimmune process. A change in the normal surface antigens of the beta cell might be caused by an exogenous factor, making these altered antigens immunogenic.

An abnormality of the immunoregulatory cells might be caused by an exogenous factor, making these altered antigens immunogenic.

An abnormality of the immunoregulatory cells might be induced by viruses, and this has been demonstrated in patients infected with congenital rubella. It has been suggested that mumps or Coxsackie viruses may directly infect the beta cells and cause diabetes, or that they become diabetogenic only against a background of a latent or potentially morbid state.

Toxins and drugs have also been incriminated in the causation of IDDM. In some cases of the so-called tropical diabetes (9), destruction and loss of islet tissue are apparent, with areas of atrophy. This disease, with its wide range of clinical and biochemical features, has been attributed to consumption of cassava (tapioca), which contains a cyanogenic glycoside toxic to the pancreas. The toxicity is aggravated by protein malnutrition, an associated condition in tropical diabetes.

The clinical manifestations of IDDM, as well as all the metabolic abnormalities, can be ascribed to insulin deficiency. The effects of the relative excess of other hormones (e.g. glucagon, growth hormone, etc.)

might also contribute to some of the metabolic features of the disease.

The absolute insulin deficiency that is the hallmark of IDDM can be clinically ascertained by measuring plasma C-peptide levels following the administration of I.V. glucagon. The lack of any excursion in the plasma levels of the C-peptide indicates total loss of insulin-secreting cells.

Treatment

The administration of exogenous insulin reverses almost all the metabolic derangements of IDDM. This treatment should be accompanied by dietary recommendations and instructions for exercise programs. It is beyond the scope of this chapter to analyze the different management regimes offered to diabetic patients. It is however important to note here several new approaches and concepts in diabetes control.

Intensive insulin treatment, including multiple daily injections of short- and long-acting insulins (MDI) or the use of the insulin pump (CSII), is recommended today. With these systems, tight control can be achieved. However, certain associated risks (e.g. hypoglycemia, infection at the site of the needle) render such intensive treatment questionable, especially in reference to the uncertainty as to whether tight control would actually prevent complications.

There is currently a preference to use highly purified insulin preparations extracted from pancreases of pigs, and also to advise certain groups of diabetics to use human insulin. The widespread distribution of the latter within the insulin-dependent diabetic community is so far hindered by cost considerations.

Another important aspect of diabetic management is the change in the dietary recommendations concerning carbohydrates. For many years, the carbohydrate intake recommended in the dietary guidelines was severely restricted. An increase of up to 50% of complexed carbohydrate is now suggested. The consumption of fiber in the diet is also being advanced, as it was found to be beneficial, and a reduced postprandial plasma excursion following pectin, guar or legume ingestion has been documented.

Since IDDM is considered to be an immune-modulated disease, a trial to use immunotherapy has been suggested. Steroids, antithymocyte globulin, plasmapheresis, levamizole and others were tested for their effect to restore beta cell function, but they were found to be inefficient or to be associated with severe adverse reactions. Recently (10), a drug called cyclosporin was tried, with some success, within six weeks of diagnosis of diabetes. This cyclic polypeptide selectively blocks the activation of T lymphocytes, and it is the main drug used today to suppress rejection of kidney and heart transplants. However, this immunosuppressive drug is not free of side effects, which include hirsutism, gingival hypertrophy, peripheral neuropathy, and most importantly, renal insufficiency and hypertension. More studies are needed to ascertain the benefits vs. the risks involved in the use of cyclosporin in young diabetic patients. It is also important to recognize the fact that when diagnosis of IDDM is made, already 90% of the beta cell mass has been destroyed. It stands to reason, therefore, that immunosuppressive therapy should be started at the preclinical stage, in order to rescue a higher proportion of insulin-secreting beta cells. The possibility to identify people at risk exists, since tests to detect ongoing beta-cell destruction (specific antibodies, lymphocyte subpopulations, etc.) are available.

Another method to restore beta cell function in chronic diabetics is pancreas transplantation. The success rate is limited, due to the failure of the immunosuppressive agents currently used both to prevent rejection and to avoid the adverse effects thereof. The procedure is restricted to patients who, simultaneously, need a kidney transplant, i.e. to those who come to the operating room with multiple organ involvement and failure. Such patients are, however, less receptive of transplants, and pancreas transplantation is still considered to be an experimental method.

NON-INSULIN-DEPENDENT DIABETES MELLITUS (NIDDM)

This disease is more prevalent than IDDM, and for every three patients with IDDM, there are 17 patients with NIDDM.

Clinical Presentation

The typical patient with NIDDM is an obese person over the age of forty, usually physically inactive and living in an urban area. Although polydipsia, polyuria and fatigue are the chief complaints in these patients, all three symptoms are less frequent, and a history of weight loss and ketosis is rare. In some cases, routine blood or urine examinations, done because of apparently unrelated symptoms, disclose the existence of the disease. Furthermore, it is more common in this type of diabetes to diagnose the disease only after the patient has already developed one or two complications.

Genetics

The transfer of NIDDM from parents to children is much more straight-forward than it is in the case of IDDM. If one of the parents has the disease, the chance of transmission to an offspring is 10-15%. The risk for a child to develop NIDDM is even higher when both parents have the disease (rate not determined, but it reaches up to 50%). In studies with twins, it has been confirmed that the concordance rate for NIDDM is close to 100%, indicating that the disease is purely genetic. However, no HLA associations have been incriminated in NIDDM, a fact which clearly distinguishes this disease from IDDM. This observation caused these syndromes to be regarded, genetically, as two totally different entities. An extra 1600 or 3400 base pairs of DNA, inserted approximately 500 base pairs upstream (5') before the transcription initiation site of the insulin gene on the eleventh chromosome, were identified using the Southern blot analysis. This length polymorphism (insertion or sometimes deletion) is more frequent in patients with NIDDM, making this system useful for predicting people at risk. Whether this polymorphism influences gene expression and insulin biosynthesis has not been established as yet.

Etiology and Pathogenesis

Being a genetically determined disease, NIDDM is not always expressed phenotypically. It would appear that there is a need for an "environmental diabetogenic" factor or an associated abnormality, in order for the disease to be clinically manifested. The finding of length polymorphism in the 5' flanking region of the human insulin gene was linked to the defective insulin biosynthesis in NIDDM. A decrease in insulin secretion as a response to various secretagogues is not sufficient cause for glucose intolerance. Moreover, the absence of the first phase of insulin secretion following glucose load, which is often the case, is not enough to explain the continuous and fasting hyperglycemia in NIDDM.

There are at least three more conditions contributing, each alone, or in concert with others, to the hyperglycemia of NIDDM.

First, an alpha cell dysfunction has been documented (11). Fasting hyperglucagonemia, despite increased plasma glucose concentrations, has been recorded in patients with NIDDM. The suppressed glucagon secretion, which occurs normally following carbohydrate ingestion in the presence or absence of increased ambient glucose, is non-operative in NIDDM. There is also an exaggerated response of alpha cells to the I.V. administration of certain amino acids in NIDDM patients.

Secondly, a role of increased hepatic glucose production in the fasting hyperglycemia of NIDDM has been proven operative (12). In the non-diabetic situation, a close-loop feedback system that includes the beta cell, the liver cell in close proximity, and the blood circulation, is functionally modulated to determine the secretion of insulin according to the increasing blood glucose levels, which then suppress hepatic glucose output, eventually decreasing blood glucose levels. In NIDDM, although there is an impaired beta cell sensitivity to glucose, these cells continue to secrete insulin. Liver cells, at the same time, are less sensitive to insulin, therefore continuing to produce more glucose to further increase blood glucose levels. This latter event only partially increases the capability of the beta cell to respond with augmented insulin secretion. This cascade of metabolic-endocrine phenomena results in sustaining the hyperglycemia in NIDDM.

Thirdly, an insulin resistance state is evident in NIDDM (13). This situation is further exaggerated by obesity. "Diabesity" is a term used today to denote a typical clinical picture of an overweight NIDDM patient. Obesity by itself is an insulin-resistant condition, as is type II diabetes. Extensive studies to characterize this peripheral resistance to insulin action provided ample data to postulate that both receptor and postreceptor defects play a role. The nature of the specific cellular components and processes responsible for the resistance to insulin action is still being debated.

The insulin resistant states are accompanied by hyperinsulinemia. This is true especially in obesity, hypercorticism and NIDDM. It has been suggested that hyperinsulinemia is the primary defect in obesity and that the receptor defect is secondary to the increased ambient insulin, due to down regulation of the insulin receptors. Hyperinsulinemia develops in an animal model (sand rat) following an increase in caloric intake and results in obesity and insulin resistance. In this model, reduced insulin uptake by the liver has been demonstrated (14). The absence of an insulin gradient concentration across the liver might be the primary event in the generation of hyperinsulinemia and the accompanying obesity. In this speculative hypothesis, obesity is secondary to hyperinsulinemia, and not the opposite, as usually assumed.

Some other abnormal situations might mimic type II diabetes. These include production of abnormal insulin molecules and mutant insulin, anti-insulin receptor antibody, primary (familial) deficiency of insulin receptors, and familial hyperproinsulinemia.

Treatment

Rather than go into details of NIDDM management, we shall review a few concepts.

Most NIDDM patients can be treated with dietary therapy only, which includes caloric restriction for the purpose of achieving weight reduction, since 80% of NIDDM patients are obese. Patients not responding to diet alone, and having continuous post-prandial hyperglycemia of more than 200 mg (11.1 mmol), should be advised to use oral hypoglycemic agents. For patients with severe hyperglycemia, a course of intensive insulin treatment is recomended, until euglycemia is achieved. Thereafter, resumption of oral hypoglycemic drug use is suggested.

The oral hypoglycemic agents in widest use today are different types of sulfonylureas (chlorpropamide, glibenclamide, and glipizide). The sulfonylureas have pancreatic as well as extra-pancreatic effects. They enhance insulin secretion by the beta cells, resulting in higher blood insulin levels, but this effect usually diminishes after several months. However, euglycemia is maintained, probably through the extra-pancreatic effects of the drug.

Another group of oral hypoglycemic drugs is the biguanides, such as phenformin and metformin. The mechanisms of action of the biguanides is by promoting the effect of insulin in peripheral tissues. A combination of insulin with metformin might reduce the requirements for insulin.

Two novel approaches to the delivery of insulin non-parenterally are now undergoing clinical trials. One is the use of a nasal spray (15) and the other the administration of rectal suppositories (16). Both methods incorporate bile salts or other non-ionic detergents to facilitate the absorption of the hormone across mucosal barriers.

COMPLICATIONS OF DIABETES MELLITUS

Hyperglycemia is the hallmark of DM, and the diagnosis and gradation of the severity of the disease are determined by the blood glucose levels during fasting and following glucose load, or after a meal. In most patients, however, several other clinical or pathological features will accompany the acute onset of the disease or will follow. These features are morbid phenomena, called complications, which are comprised of metabolic, functional, or structural disturbances. Although the nature of these acute or chronic complications is still under debate, it is reasonable to assume that they may be attributable to the metabolic derangements of the disease, and that a genetic predisposition contributes to their severity or time of appearance.

The acute complications of diabetes include diabetic ketoacidosis and nonketotic hyper-osmolar syndrome. These two acute metabolic derangements are due to absolute or relative insulin deficiency and are, therefore, correctible by the administration of the hormone. These grave and dramatic clinical syndromes are rare since the introduction of insulin therapy more than sixty years ago. Additional supportive treatment is needed in each case in accordance with the precipitating cause (e.g. infection, trauma, stress), and in accordance with the associated metabolic derangements (e.g. degree of dehydration, electrolyte deficiency, acidosis).

There are several functional disturbances in diabetes worth mentioning: patients are prone to develop infections of bacterial or fungal origin. Usually, these are focal processes in the skin, urinary tract, respiratory tract and the teeth. Various specific infections are encountered in diabetes, such as mucocutaneous candidiasis, mucormycosis of the skull, malignant otitis externa, and osteomyelitis of the bones of the feet.

A hypercoagulable state can be demonstrated in diabetes which, by itself, does not present any clinical syndrome, but might be pathogenetically important in the causation of other complications. Increased platelet aggregation and augmented blood viscosity with high levels of blood coagulation factors have been documented.

The structural changes in late complications of diabetes include neuropathy, nephropathy, retinopathy and dermopathy - referred to as microvascular disease; macrovascular complication consists of accelerated atherosclerosis, afflicting the peripheral vascular tree, the coronary arteries, and the central nervous system vasculature.

Diabetic neuropathy is one of the most common disturbances in diabetes. It is also one of the earliest to appear. The classification of the neuropathies is as follows:

1. Symmetric distal polyneuropathy.
2. Assymetric neuropathy which includes cranial mononeuropathy, peripheral mononeuropathy, and the neuromuscular syndromes.
3. Autonomic neuropathy.

The polyneuropathy presents with symmetric painful sensation, usually in the lower extremities. The pains are described as shooting or burning, occur mostly at night, and are accompanied by hyperesthesia. There is also the asymptomatic type, which is characterized by hypoesthesia and loss of deep tendon reflexes. The hypoesthesia and the insensitive feet lead to the development of foot ulcers, due to chronic pressure by tight shoes, or repeated, non-perceived traumas. This type of neuropathy is ascribed to a specific metabolic derangement that occurs in diabetes, namely the augmented polyol pathway, which culminates in an accumulated sorbitol and fructose contents in the cells. Myoinositol depletion also plays a role in this process. Following identification of these possible mechanisms, therapeutic approaches, including the use of aldose reductase inhibitors (Sorbinil) and myoinositol administration, have been employed.

The asymmetric neuropathy consists of a sudden paralysis of an isolated muscle innervated by either cranial (III, VI, IV) or peripheral nerves (footdrop, carpal tunnel syndrome). It is usually reversible and pathogenetically related to a vascular occlusion of a small nutrient arteriole by platelet thrombi.

Autonomic neuropathy includes: neurogenic bladder, gastroparesis, impotence, orthostatic hypotension. Some of these clinical presentations are disabling, distressing, and disrupting both physically and psychologically. Attempts to alleviate the symptoms of these complications are usually disappointing.

More than 85% of patients with long-standing diabetes (over 25 years) will experience changes in their kidneys and eyes. This holds true especially for patients with IDDM.

Nephropathy is a progressive functional abnormality of the glomeruli resulting in renal insufficiency. Various stages in the development of this affliction have been described. Surprisingly, hyperfunction of the kidneys can be demonstrated at the onset. Increased glomerular filtration rate with enlarged kidneys can be shown. However, within several years, signs of deteriorating renal function become apparent, with an increasing amount of proteinuria and decreased creatinine clearance. End-stage renal failure, necessitating chronic dialysis or kidney transplantation, is almost unavoidable.

The most striking and feared complication in diabetes is the loss of vision due to retinopathy. For the diabetic patient, impaired sight is serious for two reasons: he can no longer take care of himself (like measuring his own blood glucose, injecting insulin, or watching for pressure sores on his feet), and he looses the joy of life. Retinopathy has become the most common cause of blindness in civilized countries. There are two stages in the development of diabetic retinopathy: (1) Background retinopathy, and (2) Proliferative retinopathy. The first phase is rarely symptomatic but can easily be identified by fundoscopy. The second, more distressing phase, leads to impaired vision and blindness.

The non-proliferative changes include increased permeability of the capillaries seen on fluorescein angiography. Soft exudates, representing capillary non-perfusion, and hard exudates, representing leakage of lipid and protein-containing serum from the vessels, are also seen. In some cases, edema of the macula, due to leakage, can cause transient impairment of vision. Microaneurysms are almost pathognomonic to the background retinopathy in diabetes and are also seen in the fundus of the eye.

In proliferative retinopathy, new nets of capillaries grow into the vitreous from the regions in the retina that suffered from ischemic infarcts. Vasoactive growth peptides originating from the ischemic areas stimulate the proliferation of small friable vessels. Hemorrhages from these fragile vessels result in a further decrease of vision acuity. Moreover, traction on the vitreous body, whenever these hermorrhages begin to organize and fibrose, may cause retinal detachments.

New therapeutic modalities, including photocoagulation and vitrectomy, may, in many cases prevent deterioration in the condition, or even restore sight in near-blindness.

Atherosclerosis afflicting large and medium-sized arteries, especially peripheral vessels, is more common in diabetic patients than in the non-diabetic population and affects younger people (17). This macrovascular complication leads to early coronary artery disease, to episodes of strokes, and to severe peripheral vascular disease. The last-mentioned results in gangrene and loss of lower extremities. Although the clinical manifestations of atherosclerosis in diabetics are similar to those in non-diabetic patients, it seems that the medium and small vessels are more prone to involvement in diabetics than in the general population.

The most important atherogenic factors that play a role in diabetes are the increased levels of plasma total cholesterol, decreased concentrations of HDL, increased levels of certain hormones (growth hormone, catecholamines, corticosteroids, etc.), abnormalities in platelet function, and other coagulation factors. It is worth noting that the risk for diabetic patients to develop macrovascular diseases is even higher in patients with diabetic nephropathy, due to the associated hypertension and hyperlipoproteinemia.

DIABETES CONTROL AND COMPLICATIONS

The introduction of insulin treatment more than sixty, and the discovery of several hypoglycemic oral drugs some thirty years ago, have eliminated the immediate threat of death due to acute metabolic decompensation in diabetes. However, there exists some uncertainty as to whether these therapeutic modalities can totally eliminate diabetic complications. Recently, animal experiments and extensive retrospective and prospective epidemiological studies have shown some relationship between the development of vascular complications and glycemic control. The new treatment strategies for intensive insulin regimes (insulin pumps and multiple daily injections) and dietary guidelines to achieve persistent euglycemia, as well as novel methods to assess near-normal blood glucose levels (self-monitoring blood glucose and glycosylated protein measurements) are used as tools to evaluate this relationship.

The medical trend today, although not fully confirmed scientifically, is to recommend intensive treatment for newly discovered patients, designed at avoiding hypoglycemic episodes, in order to prevent late vascular sequelae. This is suggested, even though it is believed that, besides metabolic abnormalities, genetic predisposition is a contributory factor to the development of late complications.

A rough estimate (18) is that 20-25% of diabetic patients will not develop any vascular affliction, or only minimal changes, regardless of their glucose control. About 5% of patients will develop severe complications, even though they have only a slight degree of hyperglycemia. Thus, a major proportion of the diabetic population may still benefit from the attainment of near-normal blood glucose levels, in spite of various genetic predispositions in these patients. It is, however, currently impossible to classify patients according to their genetic predisposition.

ACKNOWLEDGMENTS

The author thanks the Reichmann family of Toronto, Canada, for their generous and continuous support of his research and clinical work in diabetes mellitus. The author also gratefully acknowledges the valuable assistance provided by his son Ophir in preparation and typing of the original draft of the manuscript.

REFERENCES

1. Johnston, D.G. and Alberti, K.G.M.M. (Eds.) New aspects of diabetes. In: Clinics in Endocrinology and Metabolism, Vol. 11, No. 2, W.B. Saunders Co. Ltd., London, Philadelphia, Toronto, 1982.
2. Camerini-Davalos, R.A., Velasco, C., Glasser, M. and Bloodworth, J.M.B. Drug-induced reversal of early diabetic microangiopathy. N. Engl. J. Med. 309:1551-1556, 1983.
3. Siperstein, M.D., Unger, R.H. and Madison, L.L. Studies of muscle capillary basement membrane in normal subjects, diabetic, and prediabetic patients. J. Clin. Invest. 47:1973-1999, 1968.
4. Platz, P., Jakobsen, B., Morling, N., Ryder, L.P., Svejgaard, A. *et al.*, HLA-D and -DR antigens in genetic analysis of insulin-dependent diabetes mellitus. Diabetologia 21:108-115, 1981.
5. Bell, G.I., Horita, S. and Karam, J.H. A polymorphic locus near the human insulin gene is asociated with insulin-dependent diabetes mellitus. Diabetes 33:176-183, 1984.
6. Gepts, W. Pathologic anotomy of the pancreas in juvenile diabetes mellitus. Diabetes 14:619-633, 1965.
7. Lernmark, A. Molecular biology of type I (insulin dependent) diabetes mellitus. Diabetologia 28:195-203, 1985.
8. Eisenbarth, G.S. Type I diabetes mellitus. A chronic auto-immune disease. N. Engl. J. Med. 314:1360-1368, 1986.
9. Abu-Bakare, A., Taylor, R., Gill, G.V. and Alberti, K.G.M.M. Tropical and malnutrition-related diabetes: A real syndrome. Lancet 1:1135-1138, 1986.
10. Stiller, C.R., Dupre, J., Gent, M., Jenner, M.R. and Keown, P.A. Effects of cyclosporin immunosuppression in IDDM of recent onset. Science 223:1363-1367, 1984.

11. Sussman, K. (ed.) Diabetes dialogue: The spectrum of defects in non-insulin-dependent diabetes mellitus. (Proceedings of a symposium). Am. J. Med. 79 (2B):1-44, 1985.
12. Olefsky, J.M. Introduction: pathogenesis of insulin resistance and hyperglycemia in NIDDM. Am. J. Med. 79 (3B): 1-7, 1985.
13. Kahn, C.R. Insulin resistance, insulin insensitivity, and insulin unresponsiveness. A necessary distinction. Metabolism 27 (suppl. 2):1893-1902, 1978.
14. Ziv, E., Adler, J.H., Lior, O. and Bar-On, H. Insulin uptake by the liver of sand rat. (Abstract). Diabetes 35:158A, 1986.
15. Salzman, R., Manson, J., Griffing, G.T., Kimmerle, R., Ruderman, N., McCall, A., Stoltz, E., Mullin, C., Small, D., Armstrong, J., and Melby, J.C. Intranasal aerolized insulin. N. Engl. J. Med. 312:1078-1084, 1985.
16. Raz, I., Kidron, M., Bar-On, H. and Ziv, E. Rectal administration of insulin. Israel J. Med. Sci. 20:173-175, 1984.
17. Ruderman, N.B. and Haudenschild. Diabetes as an atherogenic factor. Progr. Cardiovasc. Dis. 26:373-412, 1984.
18. Raskin, P. and Rosenstock, J. Blood glucose control and diabetic complications. Ann. Int. Med. 105:254-263, 1986.

ADDITIONAL BIBLIOGRAPHY

1. Brownlee, M. (ed.) Handbook of Diabetes Mellitus (in 5 volumes). John Wiley and Sons, New York, 1981.
2. Keen , H. and Jarret, J. (eds.) Complications of Diabetes (2nd edition). Edward Arnold, London, 1982.
3. Ellenberg, M. and Rifkin, H. (eds.) Diabetes Mellitus, Theory and Practice (3rd ed.). Medical Examination Publishing Co. New Hyde Park, NY, 1983.
4. Alberti, K.G.M.M. and Krall, L.P. (eds.) The Diabetes Annual/1. Elsevier Science Publishers, New York, 1985.
5. Olefsky, J.M. and Sherwin, R.S. (eds.) Diabetes Mellitus: Management and Complications. Vol. 1 of "Contemporary Issues in Endocrinology and Metabolism, Churchill Livingstone Inc., New York, 1985.

6

AUTOIMMUNITY IN DIABETES

H. MARKHOLST and Å. LERNMARK, Hagedorn Research Laboratory, Niels Steensensvej 6, DK-2820 Gentofte, Denmark

INTRODUCTION

Before the turn of the century, insulin-independent diabetes mellitus (IDDM) was shown to develop in pancreatectomized dogs (1), thus linking the pathogenesis of the syndrome to the pancreas. Later, insulin was extracted from the pancreas and proved to ameliorate most of the diabetic symptoms(2). However, it became clear that insulin was not a cure. In 1965, Gepts reported (3), that the major morphologic alteration in the pancreas at the onset of IDDM was disruption of the architecture of the islets of Langerhans and a loss of cells. The number of B cells was reduced to less than 10% of normal values and in 68% of the pancreata examined, the islets were infiltrated by mononuclear cells. Islet inflammation, insulitis, is therefore thought to be of pathogenetic importance (3-6). Infiltration of the islets of Langerhans has also been associated with experimental diabetes introduced by immunization with insulin (7-8), by multiple injections of low dose streptozotocin (9) or following infection with diabetogenic viruses (10,11).

Further support to the hypothesis that IDDM has an autoimmune pathogenesis was obtained from the demonstration of a hypersensitivity reaction to pancreatic antigens (12), islet cell antibodies (13,14) and a close association to the HLA locus on chromosome 6 (15-17). The pathogenesis of IDDM may therefore be due to an abnormal immune response directed against the pancreatic B cells, causing a specific loss of these cells and a resulting diabetes. In this chapter some of the autoimmune relationships to IDDM will be put into perspective. In view of the rapidly growing literature in this field our synthesis will be selective, and the reader, therefore, is referred to additional excellent reviews (18-20).

The normal immune response

The pathway of cellular interactions involved when the immune system mounts a response against a foreign antigen (fAg) is complex and not yet fully understood (Fig.1). This pathway is also assumed to be involved in a reaction against "self antigens" or autoantigens (aAg). In brief, a fAg is taken up by antigen-presenting cells (APC), which are macrophages or macrophage-like cells. The APC process the fAg, and are thought to present an epitope of the Ag on its surface in conjunction with or perhaps even bound to a Class II molecule. The latter is encoded in the major histocompatibility complex (MHC) or HLA on the short arm of the human chromosome 6.

Becker, Y (ed), Virus Infections and Diabetes Melitus. © 1987 Martinus Nijhoff Publishing, Boston. ISBN 0-89838-970-4.

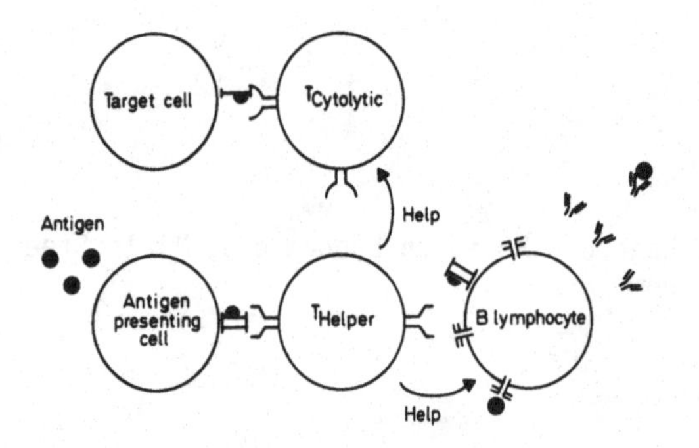

FIGURE 1. Schematic representation of pathways of an immune response to a foreign antigen.

A T-helper lymphocyte expressing a specific T-cell receptor is activated, when the Ag-Class II complex is recognized. The activation of T-helper lymphocytes provides stimulus to self proliferation and causes clonal expansion and differentiation of effector cells.

Effector cells are Ag-specific T lymphocytes with either cytotoxic or suppressive properties and/or B lymphocytes. Cytotoxic T lymphocytes are able to recognize and kill target cells which express the particular fAg epitope on their cell surface. However, the killing is restricted to target cells which share Class I molecules (HLA-ABC region gene products) with the cytotoxic T lymphocyte. B lymphocytes express membrane-bound antibody molecules, which serve as Ag receptors.

These bind circulating fAg, but they will only proliferate and differentiate into plasma cells when they receive T cell help (i.e. growth factors, lymphokines etc). B lymphocytes have also been shown to process and present Ag. Plasma cells secrete large amounts of antibody (Ab) that combine to circulating fAg to form complexes. Recent evidence suggests that each clone of antigen-activated B lymphocytes experiences rapid somatic mutations at the rate of 10^{-3} per base pair per cell division (21) thus modifying the immunoglobin genes to provide an increased Ag specificity.

The effector mechanism of the Ab includes binding to an invading organism. This binding results in an effective activation of the complement pathway. The end result is lysis of the organism. Similarly, an Ab may bind to the surface of a cell thereby eliminating this cell by the same mechanism. Antibodies may also arm killer cells (leukocytes, macrophages) to seek a target Ag on a cell surface to mediate an antibody-dependent cellular cytotoxic reaction. Finally, an antibody may bind to a specific receptor either to mimic the receptor ligand or to act as an antagonist.

The formation of Ab seems controlled by the availability of Ag, but also in part by feedback mechanisms. The latter phenomenon involves the formation of Ab which recognize the antigen-binding portion of the original Ab. This portion of the Ab is the idiotype and the Ab formed against the idiotype is termed anti-idiotypic Ab. Such Ab are thought to be important effectors of control (22), but have also been implied in the development of autoimmune disease (23). The clonal expansion of effector cell populations may, furthermore, be controlled by T suppressor lymphocytes or vetocells.

The autoimmune response

It is assumed that the abovementioned cellular interactions are involved in an immune response which develops against an autoantigen (aAg). A fAg may induce Ab which crossreact with an aAg. A well-known example of this is the occasional occurrence of hemolytic episodes during mycoplasma pneumonia infections, in which Ab known as cold agglutinins also react with the group I receptor of erythrocytes to cause complement-dependent hemolysis (24). Infection with group A streptococci may cause rheumatic fever in an HLA-DR associated manner to involve autoreactivity against cardiac valve receptors (25). The Yersinia enterocolitica membrane possesses a saturable binding site (Ag-epitope) for the thyroid-stimulating hormone to mimic the thyroid cell TSH receptor, which may explain why epidemics of Yersinia are followed by an increased incidence of Graves disease (26).

These pathways may be circumvented by factors which induce a polyclonal activation. Such activators are naturally-occurring. Epstein-Barr virus (EBV) which causes infectious mononucleosis may induce Ab directed against various aAg present in diverse tissues and cells such as smooth muscle, nuclear proteins, lymphocytes and erythrocytes (27). Parasitic (28) or protozoan (29) infections have also been shown to cause unspecific B lymphocyte stimulation.

Several other mechanisms of a deterioration of the natural tolerance to aAg may be listed. A phenomenon which still requires an explanation, is the by now classical associations between certain HLA types and autoimmune diseases - the hypothesis is that genes encoded in the MHC confer susceptibility to certain autoimmune reactions. However, the mechanisms involved remain to be clarified.

If autoimmunity is important in the etiology and pathogenesis of diabetes, there is inadequate information concerning the initial steps of how the islet B-cell becomes the specific target of the immune system.

Genetic susceptibility

IDDM is known to run in families, but the hereditary factor(s) is not sufficient, since the concordance in monozygotic twins is 50% or less (19-20). Tissue typing has revealed that 98% of all Caucasian IDDM patients are DR 3 and/or 4 positive (15-16). However, the incidence of these particular genotypes in the general population is high, nearly 60%. The DR 3 and/or 4 association seems to be explained by the linkage of the serologically determined DR-specificities to Class II antigens encoded for by the D-region in the HLA (Fig. 2).

FIGURE 2.

The Class II antigens consist of two dissimilar subunits called α and β chains being M_r 34000 and M_r 29000, respectively. Each D-region locus encodes one or several α or β chains. A cloned cDNA of one of the HLA-DQ β chain genes was used in a restriction fragment length polymorphism (RFLP) analysis to test, whether HLA-DR identical IDDM and control individuals were different at certain restriction enzyme sites (17). The DR4 haplotype was found (30,31) to be linked to a BamHI restriction site to generate a 3,7 kb fragment, which compared to the controls was reduced among the diabetic patients (17). This observation has later been confirmed by others (32,33). The BamHI restriction sites were found to be located in intron sequences of an HLA-DQ β chain gene (34). Further analyses suggest that the HLA-DQ β chain locus is closer to a diabetogenic locus than is HLA-DR (35). Complete molecular cloning of HLA-D region genes will provide gene probes (36) that will be correlated with available serological tissue typing reagents to define better the gene product(s) which provide susceptibility to IDDM. It is still not proven whether the Class II molecules are involved in the pathogenesis of IDDM. It cannot be excluded that yet another gene, or genes, in linkage equilibrium with the Class II molecule genes encodes the factual diabetogenic activity.

It is often speculated that several susceptibility genes are needed for IDDM to develop. Various blood group markers have been studied but none of them have shown a close and reproducible association to IDDM (37). The mode of inheritance is therefore still unclear. It should be taken into account that genes on chromosomes other than chromosome 6 may be important to promote a diabetogenic activity. Again, it is striking that about 50% of monozygotic twins remain discordant for IDDM (38) and that only 13% of children (0-16 years) developing IDDM, in Sweden, have a sibling or parent with the disease (39). It is therefore generally assumed that yet unknown environmental factors initiate processes that eventually allow IDDM to develop.

The discovery of the spontaneously diabetic BB rat (40,41) and attempts to breed these animals properly, to study the genetic inheritance, have provided some evidence for the possibility that a diabetes gene may be located outside the major histocompatibility complex. The spontaneously diabetic BB rat presents a diabetes which is essentially inseparable from the human counterpart (41).

The diabetic BB rat is of the MHC $RT1^u$ background and breeding studies with congenic lines suggest that the $RT1^u$ haplotype in the B,D, and/or C/E loci, but not the A, are necessary for diabetes to develop (42). RFLP analysis of Class I and II genes have failed to give conclusive results of a genetic polymorphism associated to IDDM (43,44). In our own laboratory, Class II antigen gene probes have not shown a polymorphism between our diabetic and diabetes-resistant rats (45). A Class I gene in one of the diabetes-prone BB rat lines was found to be dissociated from diabetes, but linked to a plasma protein able to bind β_2 microglobulin (45) and therefore thought to be a Class I molecule. Although it cannot be excluded that minor differences in Class I and II genes exist between otherwise $RT1^u$ identical non-diabetic and diabetic rats, there is evidence to suggest that a susceptibility gene(s) may reside outside the MHC (44).

In summary, the observed tendency to develop IDDM in HLA-DR3 and/or 4 positive individuals may be explained by the close association to the HLA-DQ locus. The gene(s) located within the HLA complex may control only IDDM susceptibility, as the concordance among monozygotic twins is less than 50% and less than 30% among HLA identical siblings. It is therefore hypothesized that environmental factors such as viruses or chemical agents are factors necessary to initiate an abnormal immune response which eventually is directed towards the islet B cells. The role of an environmental factor, such as a virus, may explain observed differences in the incidence of IDDM with respect to season (46) and geographical location (39).

Islet B-cell autoantigens

A major problem in understanding IDDM is to explain the specific loss of the insulin-producing cells. This would be easily done, if antigen-specific effectors, either CTL or antibodies could furnish the immune system with a mechanism by which to detect only the pancreatic B cells. What particular molecule of the B cell may therefore be the target of immune destruction? Since all islet B cells, including those transplanted, are erradicated in diabetic BB rats (47), the target is likely to be a molecule specifically and normally expressed on the B cells. The approach to detect islet autoantigens has been to label islet proteins by metabolic labelling *in vitro* and then immuneprecipitate the labelled proteins with sera from BB rats (48) or subjects with (49) or without (50) IDDM. Ab-Ag complexes formed are absorbed to Protein A-Sepharose, washed and boiled in SDS to denature and solubilize bound Ab-Ag. SDS gel electrophoresis of the solubilized material separates the proteins according to their electrophoretic mobility. It was detected that autoantibodies in diabetic sera specifically recognized a protein of about M_r 64000 (49). The diabetic sera were also able to bind a rat islet B-cell protein of similar molecular mass, while this protein could not be detected in non-islet tissues (51).

So far, this is the only evidence of possible islet B-cell-specific Ag which might be involved in the pathogenesis of IDDM. Since the pancreatic islet cells also synthesize and express Class I antigens (52,53), they may serve as targets of T-cytotoxic (CTL) killing, if the proper Ag is presented and seen by the CTL receptor. The nature of the M_r 64000 protein remains obscure, except that it is a membrane-associated molecule and may be present in minor concentrations on the cell surface of human and rat islet cells. Whether it is a receptor molecule of importance to specific properties of the B cells such as glucose-stimulated insulin biosynthesis, storage and release remain to be determined.

How is the autoantigen presented?

It is assumed that the formation of autoantibodies against the islet B cells is preceded by antigen processing,presentation and activation of the B lymphocyte circuit, by a proper T helper lymphocyte.

It is still unclear to what extent the islet B cell itself may serve as an APC. Given an exogeneous challenge such as a viral infection, the B cell might express Class II antigens and perhaps itself present the M_r 64000 autoantigen. Although some Class II molecule positive islet cells were reported in one subject (5), diabetic BB rats failed to express Class II molecules in the enclosing islet cells (54). It is therefore equally plausible that the initial antigen presentation causing later B-cell intolerance is initiated away from the islets of Langerhans, and that the immune recognition of the pancreatic B-cells occurs later on.

Effectors of islet B-cell destruction

Several effector mechanisms of B-cell destruction may be listed (Table 1). *In vitro* experiments are still the approaches taken to investigate possible mechanisms by which the pancreatic B cells are specifically destroyed. It is noted that passive transfer of IDDM from man to experimental animals has not been accomplished.

TABLE 1. Possible immune effector mechanisms to destroy the islets of Langerhans.

Effector	Abbreviation	Effects on the islet B cell
Islet cell surface antibodies	ICSA	May inhibit glucose-stimulated insulin biosynthesis and release
Cytotoxic islet cell antibodies	C'AMC	May mediate complement-dependent cytotoxicity
Islet cell surface antibodies	ADCC	May mediate antibody-dependent cellular cytotoxicity
Cytotoxic T lymphocytes	CTL	May kill islet B cells due to autoreactivity with an islet cell antigen
NK cells		May kill islet B cells
Islet macrophages		May kill islet B cells by IL-1 cytotoxicity

B lymphocytes

The variable portion of the immunoglobulin molecule - that is the antigen-binding site - is coded for by separate genes for the heavy and light chains. These genes are present in germline DNA but following antigen stimulation, the DNA is not only rearranged in the somatic DNA but the genes for the variable sequences are also the subject of an extensive degree of mutations.

The net result is highly diverse Ab with different affinities for the Ag and also variable ability to mediate complement-dependent cytotoxicity. The detection of Ab bound to its Ag is therefore highly dependent on the assay employed. The large number of techniques, which have been used, makes it therefore difficult to compare results from different laboratories. Little is known also about blood B lymphocytes that actively produce antibodies against islet cell Ag. Provided the Ag is available to coat an indicator red blood cell, it is possible to use a plaque assay (55) to determine the number of B lymphocytes producing antibodies against the antigen. Although such experiments remain to be carried out in IDDM with islet B-cell-specific B lymphocytes, it has been reported that the number of B lymphocyte plaques spontaneously producing antibodies are increased in newly diagnosed IDDM patients (56,57).

Islet cell antibodies (ICA). Islet cell (cytoplasmic) Ab (ICA) is detectable by indirect immunofluorescence using frozen sections of human pancreas (5,13). The ICA are not B-cell specific, as they bind to the cytoplasmic determinants in all islet cells, which is evident when double-antibody analysis is carried out with insulin or proinsulin antibodies (58). The aAg have not been identified. The significance of the ICA in the pathogenesis of IDDM has been questioned, since the Ab binding is intracellular and affects all islet cells.

At the clinical onset, ICA are detected in 60-80% of the patients (5,13), but recent studies indicate that ICA may also be abundant prior to the initiation of insulin therapy (59,60). The ICA disappear with the increasing duration of IDDM, in most patients. The rate of disappearance was correlated to the titer at the time of clinical diagnosis. (61).

In addition, the decline in C-peptide, perhaps reflecting the number of islet B cells, was found to be associated to the ICA titers (61). The levels of ICA in relation to residual B-cell function should be carefully evaluated, since in one study, the ICA levels decreased rapidly during immunosuppressive therapy (62). It is of interest that ICA have not been detected in the spontaneously diabetic BB rat (20).

Islet cell surface antibodies (ICSA). Autoantibodies directed against the B-cell surface (ICSA) were detected by incubating dispersed viable rat islet cells with diabetic sera. After washing, the cells were exposed to a FITC-labelled second antibody. Bound aAb were visualized in either the light microscope or the FACS (fluorescence-activated cell sorter) (14,63).

Alternatively, the ICSA are detected in a radioimmunoassay using e.g. ^{125}I labelled Protein A (64), which binds specifically to the F_c of most, but not all, IgG subclasses. The latter assay may therefore not detect all ICSA, but offers the advantage of being quantitative and sensitive.

The ICSA appear to be B-cell specific at least in IDDM patients that are younger than 20 years at onset (63). In one series of sera, there was a good correlation between ICSA and antibodies against the M_r 64000 protein (49). The ICSA are readily detected at onset and disappear after onset, suggestive of a marker of B-cell intolerance (14). The prevalence of ICSA is lower than that of ICA, and comparative studies have shown that only about 50% of IDDM sera at the time of clinical onset were both ICA and ICSA positive.

The ICSA have been detected in the sera of BB rats well ahead of the time when insulitis is detectable. So far, M_r 64000 autoreactivity has been detected already at the time of weaning, which is 60 to 90 days prior to IDDM (48) - while the ^{125}I Protein A radioligand assay detected ICSA about 20 days later (65). Using cell suspensions of purified B cells and non-B cells, a recent investigation demonstrates that BB rats 100 days or younger at the time of clinical diagnosis have B-cell specific ICSA (66). The B-cell specific ICSA were also found to mediate complement-dependent cytotoxicity. A schematized time course of immune phenomena in the BB rat (Fig.3) demonstrates the current view of the natural history of the pathogenesis in these rats. The passive role of these islet cell surface autoantibodies in the initial cell intolerance is of particular interest. ICSA positive serum was shown to mediate (in vitro) cytotoxicity to cultured B cells in the presence of complement (67).

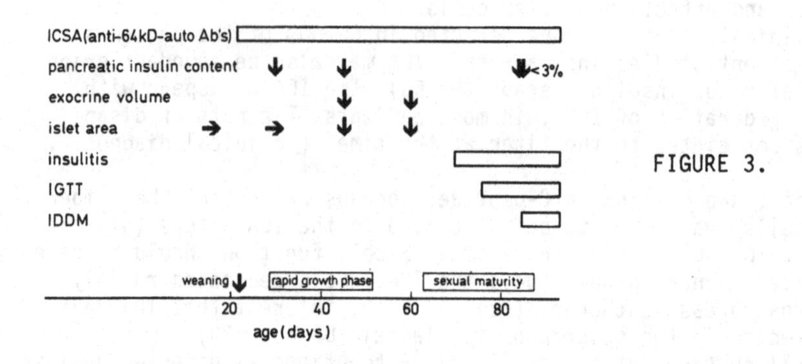

FIGURE 3.

Cytotoxic antibodies, however, do not seem restricted to IDDM patients, since they were found among 30% of first degree relatives (68). Recently, siblings to IDDM children with ICSA were shown to have elevated levels of fasting serum proinsulin, which suggests that these aAb are associated with an altered B-cell function (69). Maybe, a suppressive mechanism protects the B cell in vivo in the relatives or the cytotoxicity in vivo needs the cooperation of other cells, (i.e. killer cells), to exert their cytotoxicity. The effects of ICSA could, however, be other than mediating cytotoxicity, and it cannot be excluded that ICSA interact with molecules on the B-cell surface, which are important to the B-cell function. Rat islet cells were found to synthesize less (pro-)insulin, when incubated with fractionated IDDM serum containing ICSA (70), and in vitro perifusion in the presence of ICSA-containing media revealed a lower insulin secretion following glucose-stimulated insulin release (51).

Organ-specific autoantibodies
Apart from Ab directed against islet cells, many studies (18-20) have shown the presence of a number of aAb against various tissues such as thyroid, gastric mucosa and adrenal. This break of tolerance, which seems to affect several endocrine tissues, is not understood. Also in the BB rat (71,72) there is a similar high prevalence of organ-specific aAb. In newly diagnosed IDDM patients circulating immune complexes were detected among 25% as compared to 7% of controls (73). Whether these complexes reflect an increased autoimmunity or are of pathological significance remains to be clarified.

T lymphocytes
Effector cells of T-cell origin may also be B-cell specific. Evidence of activated or sensitized lymphocytes was obtained in tests of leucocyte migration inhibition, demonstrating that IDDM patients were sensitized against pancreatic antigens (12). Diabetic lymphocytes have been found to be toxic and to inhibit insulin release in vitro of rat islets (74).

T-suppressor lymphocytes may be decreased at onset, while NK cells may either be increased (75) or decreased (76) dependent on the methods used to detect these cells. NK-like cells with specificity for rat islet cells were detected in the spontaneously diabetic BB rat (77). Reports of induction or acceleration of IDDM in diabetes-prone BB rats following administration of Concanavalin A-stimulated spleen cells obtained from acutely diabetic BB rats raises the question of the pathogenetic role of cell-mediated B-cell destruction (78,79). Irradiation of these cells failed to induce this acceleration of IDDM, however, it also failed to prevent later development of IDDM (80).

Modulation of effector cell polulations

IDDM may be prevented by various means of effector cell deletion or by enhancement of protective factors (Table 2). Since these experiments are not immediately applicable to man and most immune suppressive drugs have serious side effects, our summary primarily dealswith the BB rat.

Cyclosporin A is an effective drug to prevent the spontaneous diabetes in this animal. Early initiation of therapy is necessary (81). Thymectomy, bone marrow transplantation and blood transfusions are all drastic approaches of therapy. The conclusions from these experiments are that the immune system is important to the development of IDDM in the BB rat. However, specific immunosuppressive treatment directed against those lymphocyte clones, which precisely direct their activity against the islet B cells, needs to be developed.

TABLE 2. Methods of prevention of IDDM.

Method	Recipient		Effect
Cyclosporin A	man (recent onset)	(81)	some, often transient remission
Cyclosporin A	BB rat a) diabetes prone		
	age < 40 days	(82)	100% permanent prevention
	> 40 days	(83)	less than 100% prevention often transient remission
	b) diabetes resistent	(82)	up to 60% diabetes induction
Thymectomy (nn)	BB rat	(84)	high protection
Bonemarrow transplantation of:			
RT 1 u genotype	BB rat	(85)	high protection
RT 1 1 genotype	BB rat	(86)	no protection
Blood transfusion, lymphocyte-transfer	BB rat	(87)	some prevention - lack of follow-up

Conclusion

The following phenomena relevant to a possible role of autoimmunity in IDDM are:

1) A prolonged period of apparent B cell autoreactivity precedes onset of IDDM;
2) Both humoral and cellular immune effectors may be potential killers of islet B cells;
3) Early immune intervention may prevent IDDM.

Crucial information on the mechanisms, by which the B cells specifically are lost, are still needed. Mass isolation and characterization of specific B-cell antigens is central to the understanding of the pathogenesis and etiology of IDDM. Once such antigen(s) become available, various means exist to obtain a reliable marker of ongoing B-cell attack and thus restore the tolerance to it by means of specific deletion of critical lymphocyte subsets.

REFERENCES

1. Mering J von, Minkowski O: Diabetes Mellitus nach Pankreasextirpation: Arch. Exp. Pathol. Pharmakol. 26: 371-387, 1889.
2. Banting FG, Best CB, Collip JB, Cambell WR, Fletcher AA: Pancreatic extract in the treatment of diabetes mellitus: Canad. Med. Ass. J. 2: 141-146, 1922.
3. Gepts W: Pathologic anatomy of the pancreas in juvenile diabetes mellitus: Diabetes 14: 619-633, 1965.
4. Foulis AK, Stewart JA: The pancreas in recent-onset Type 1 (insulin-dependent) diabetes mellitus: insulin content of islets, insulitis and associated changes in the exocrine acinar tissue: Diabetologia 26: 456-461, 1984.
5. Bottazzo GF, Dean BM, McNally JM, McKay EH, Swift PGF, Gamble DR: In situ characterization of autoimmune phenomena and expression of HLA molecules in the pancreas in diabetic insulitis: New Engl. J. Med. 313: 353-360, 1985.
6. Rahier J, Goebbels RM, Henquin JC: Cellular composition of the human diabetic pancreas: Diabetologia 24: 366-371, 1983.
7. Lacy PE, Wright PH, Silverman JL: Eosinophilic infiltration in the pancreas of rats injected with anti-insulin serum: Fed. Proc. 22: 60, 1963.
8. Renold AE, Soeldner JS, Steincke J: Immunologic studies with homologous and heterologous pancreatic insulin in the cow: Ciba Foundations Colloquia 15: 122, 1964.
9. Like AA, Rossine AA: Streptozotocin - induced pancreatic insulitis: new model of diabetes mellitus: Science 193: 415-417, 1976.
10. Craighead JE: The role of viruses in the pathogenesis of pancreatic disease and diabetes mellitus: Prog. Med. Virol. 19: 161-214, 1975.
11. Notkins AL: Virus-induced diabetes mellitus: Brief review: Arch. Virol. 54: 1-17, 1977.
12. Nerup J, Andersen OO, Bendixen G, Egeberg J, Poulsen JE: Anti-pancreatic, cellular hypersensitivity in diabetes mellitus. Antigenic activity of fetal calf pancreas and correlation with clinical type of diabetes: Acta Allergol 28: 223-230, 1973.
13. Bottazzo GF, Florin-Christensen A, Doniach D: Islet cell antibodies in diabetes mellitus with autoimmune polyendocrine defiencies: Lancet ii: 1279-1283, 1974.
14. Lernmark Å, Freedman ZR, Hofmann C, Rubenstein AH, Steiner DF, Jackson RL, Winter RJ, Traisman HS: Islet cell surface antibodies in juvenile diabetes mellitus: N. Engl. J. Med. 299: 375-380, 1978.
15. Platz P, Jackson BK, Morling M, Ryder LP, Svejgaard A, Thomsen M, Christy M, Kromann H, Benn J, Nerup J, Green A, Hauge M: HLA-D and DR-antigens in genetic analysis of insulin-dependent diabetes mellitus: Diabetologia 21: 100-115, 1981.
16. Barbosa J, Chavers B, Dunsworth T, Michael A: Islet cell antibodies and histocompatibility antigens (HLA) in insulin-dependent diabetes and their first degree relatives: Diabetes 31: 585-588, 1982.
17. Owerbach D, Lernmark Å, Platz P, Ryder LP, Rask L, Peterson PA,

Ludvigson J: HLA-D region β-chain DNA endonuclease fragments differ between HLA-DR identical healthy and insulin-dependent diabetic individuals: Nature 303: 815-817, 1983.
18. Andreani D, DiMario U, Federlin KF, Heding LG (eds): Immunulogy in Diabetes: Kimpton Medical Publications, London, 1984.
19. Todd I, Bottazzo GF: Laboratory investigation of autoimmune endocrine diseases: Clinics in Immunol. and Allergy 5: 613-638, 1985.
20. Rossini AA, Mordes JP, Like AA: Immunology of insulin-dependent diabetes mellitus: Ann. Rev. Immunol. 3: 289-320, 1985.
21. Honjo T: Origin of immune diversity: genetic variation and selection: Ann. Rev. Biochem. 54: 803-830, 1985.
22. Jerne NK: Towards a network theory of the immune system: Ann. Immunal. (Paris) 125: 373-382, 1974.
23. Wasserman NH, Penn AS, Freimuth PI, Treptow N, Wentzel S, Cleveland WL and Erlanger BF: Anti-idiotypic route to anti-acetylcholine receptor antibodies and experimental myasthenia gravis: Proc. Natl. Acad. Sci. USA 79: 4810-4816, 1982.
24. McConnell I, Munro A, Waldmann H: Autoimmune Disease in the immune system - a course on the molecular and cellular basis of immunity, 2nd ED: Blackwell Scientific Publ. London, p. 273, 1981.
25. Ayoub EM, Barret DJ, MacLaren NK, Krischer JP: Association of Class II Human Histocompatibility Leykocyte antigens with rheumatic fever: J. Clin. Invest. 77: 2019-2026, 1986.
26. Weiss M, Ingbar SH, Winblad S, Kaspar DL: Demonstration of a saturable binding site for thyrotropin in Yersinia enterocolitica: Science 219: 1331-1333, 1983.
27. Waldmann TA, Broder S: Polyclonal B-cell activators in the study of the regulation of immunoglobulin synthesis in the human system: Adv. Immunol. 48: 196-200, 1982.
28. Walls KW, Smith JW: Serology of parasite infections: Int. J. Parasitol. 10: 329-340, 1979.
29. Perrin LH, Mackey LJ, Lambert PH: Immunology of malaria: In: Clinical Immunal. Update (Rosse WF, ed) Elsevier, N.Y. pp. 234-264, 1985.
30. Owerbach D, Lernmark Å, Rask L, Peterson PA, Platz P, Svejgaard A: Detection of HLA-D/DR-related DNA polymorphism in HLA-D homozygous typing cells: Proc. Natl. Acad. Sci. USA 80: 3758-3761, 1983.
31. Owerbach D, Hägglöff B, Lernmark Å, Holmgren G: Susceptibility to insulin-dependent diabetes defined by restriction enzyme polymorphism of HLA-D region genomic DNA: Diabetes 33: 958-965, 1984.
32. Cohen-Haguenauer O, Robbins E, Massart C, Busson M, Deschamps I, Hors J, Lalouel JM, Dausset J, Cohen D: A systematic study of HLA class II-β DNA restriction fragments in insulin-dependent diabetes mellitus: Proc. Natl. Acad. Sci. USA, 82: 3335-3339, 1985.
33. Böhme J, Carlsson B, Wallin J, Möller E, Persson B, Peterson PA, Rask L: Only one of each DR specificity is associated with insulin-dependent diabetes: J. Immunol. 137: 941-947, 1986.
34. Michelsen B, Kastern W, Lernmark Å, Owerbach D: Identification of an HLA-DC β-chain related genomic sequence associated with insulin-dependent diabetes: Biomed. Biochim. Acta 44, 1: 33-36, 1985.
35. Michelsen B, Lernmark Å: HLA-DQ gene polymorphism in insulin-dependent diabetes: Proc. of the Satelite Symposium Immunology in Diabetes, Edmonton, in press, 1986.
36. Boss JM, Strominger JL: Cloning and sequence analysis of the human major histocompatibility complex gene DC -β3: Proc. Natl. Acad. Sci. USA. 81: 5199-5203, 1985.

37. Rotter JI, Rimoin DL: The genetics of the glucose intolerance disorders: Am. J. Med. 70: 116-126, 1981.
38. Barnett AH, Eff C, Leslie RDG, Pyke DA: Diabetes in identical twins: a study of 200 pairs: Diabetologia 20: 87-93, 1981.
39. Dahlquist G, Blom L, Holmgren G, Hägglöf B, Larson Y, Sterky G, Wall S: The epidemiology of diabetes in Swedish Children 0-14 years - A six-year prospective study. Diabetologia 28: 802-808, 1985.
40. Nakhooda AF, Like AA, Chappel CI, Murray FT, Marliss EB: The spontaneously diabetic Wistar rat. Metabolic and morphologic studies: Diabetes 26: 100-112, 1977.
41. The juvenile Diabetes Foundation workshop on the spontaneously diabetic BB rat: Its potential for insight into human juvenile diabetes: Marliss EB (ed): Metabolism, 32 suppl 1: 1-166, 1983.
42. Ono SJ, Colle E, Guttmann RD, Fuks A: MHC association of IDDM in the BB rat maps to permissive immune response genes. In "Immunology of diabetes", Molnar GD, Jaworski MA, eds: Elsevier publ., New York, (in press), 1986.
43. Jackson RA, Buse JB, Rifai R, Pelletier D, Milford EL, Carpenter CB, Eisenbarth GS: Diabetes in BB rats: J. Exp. Med. 159: 1629-1636, 1984.
44. Kastern W, Dyrberg T, Schöller J, Kryspin-Sørensen I: Restriction fragment polymorphisms in the major histocompatibility complex of diabetic BB rats: Diabetes 33: 807-809, 1984.
45. Björck L, Kryspin-Sørensen I, Dyrberg T, Lernmark Å, Kastern W:A deletion in a rat major histocompatibility complex class I gene is linked to the absence of β_2-microglobulin - containing serum molecules: Proc. Natl. Acad. Sci. USA (in press), 1986.
46. Christau B, Kromann H, Ortved Andersen O, Christy M, Bushard K, Armung K: Seasonal and geographical patterns of juvenile-onset insulin-dependent diabetes in Denmark: Diabetologia 13: 281-284, 1977.
47. Weringer EJ, Like AA: Immune attack on pancreatic islet transplant in the spontaneously diabetic Biobreeding/Worcester (BB/W) rat is not MHC restricted: J. Immunol. 134: 2383-2386, 1985.
48. Bækkeskov S, Dyrberg T, Lernmark Å: Autoantibodies against a M_r 64000 islet cellprotein precede the onset of spontaneous diabetes: Science 224: 1348-1350, 1984.
49. Bækkeskov S, Nielsen JH, Marner B, Bilde T, Ludvigson J, Lernmark Å: Autoantibodies in newly diagnosed diabetic children immunoprecipitate human pancreatic islet cell proteins: Nature 298: 167-169, 1982.
50. Bækkeskov S, Bruining S, Srikanta T, Mandrup-Poulsen T, Beaufort C, Eisenbarth GS, Lernmark Å: Antibodies to a M_r 64000 human islet cell protein in the prediabetic period of IDDM patients: Ann. N.Y. Acad. Sci. 475: 415-417, 1986.
51. Kanatsuna T, Bækkeskov S, Lernmark Å, Ludvigson J: Immunoglobulin from insulin-dependent diabetic children inhibits glucose-induced insulin release: Diabetes 32: 520-524, 1983.
52. Bækkeskov S, Kanatsuna T, Klareskog L, Nielsen DA, Peterson PA, Rubenstein AH, Steiner DF, Lernmark Å: Expression of major histocompatibility antigens on pancreatic islet cells: Proc. Natl. Acad. Sci. USA 78: 6456-6460, 1981.
53. Faustman D, Hauptfeld V, Davie JM, Lacy PE, Schreffler DC: Murine pancreatic-cells express H-2K and H-2D but not Ia antigens: J. Exp. Med. 151: 1563-1568, 1980.
54. Dean BM, Walker R, Bone AJ, Baird JD, Cooke A: Pre-diabetes in the spontaneously diabetic BB/E rat: lymphocyte subpopulations in the pancreatic infiltrate and expression of MHC class II molecules in endo-

crine cells: Diabetologia 28: 464-466, 1985.
55. Agger R, Petersen J, Dinesen B, Wiik A, Andersen A: Production and secretion of immunolobulins by in vitro activated lymphocytes: Allergy 37: 179-185, 1982.
56. Horita M, Suzuki H, Onodera T, Ginsberg-Fellner F, Fanci AS, Notkins AL: Abnormalities of immunoregulatory T cell subsets in patients with insulin-dependent diabetes mellitus: J. Immunol 129: 1426-1429, 1982.
57. Papadopoulos G, Petersen J, Andersen V, Lernmark Å, Marner B, Nerup J, Binder C: Spontaneous in vitro immunoglobulin secretion at the diagnosis of insulin-dependent diabetes: Acta endocrinol. 105: 521-527, 1984.
58. Madsen OD, Landin-Olsson M, Bille G, Sundkvist G, Lernmark Å, Dahlquist G, Ludvigson J: A two-colour immunofluorescence test with 1 monoclonal human proinsulin antibody improves the assay for islet cell antibodies: Diabetologia 29: 115-118, 1986.
59. Gorsuch AN, Spencer KM, Lister J, McNally JM, Dean BM, Bottazzo GF, Cudworth AG: The natural history of type 1 (insulin-dependent) diabetes mellitus: Evidence for a prolonged prediabetic period: Lancet ii: 1363-1365, 1981.
60. Srikanta S, Gimda O, Eisenbarth G, Soeldner JS: Islet cell antibodies and beta cell function in monozygotic triplets and twins initially discordant for type 1 diabetes mellitus: N. Engl. J. Med. 308: 322-325, 1983.
61. Marner B, Agner T, Binder C, Lernmark Å, Nerup J, Mandrup-Poulsen T, Walldorf S: Increased reduction in fasting C-peptide is associated with islet cell antibodies in Type 1 (insulin-dependent) diabetic patients: Diabetologia 28: 875-880, 1985.
62. Mandrup-Poulsen T, Nerup J, Stiller CR, Marner B, Bille G, Heinrichs D, Marstell R, Dupse J, Keown PA, Jenner MR, Rodger NW, Wolfe B, Graffenried BV, Binder C: Disappearance and reappearance of islet cell cytoplasmic antibodies in cyclosporin - treated insulin-dependent diabetics: Lancet i: 599-602, 1985.
63. Winkel van de M, Smets G, Gepts W, Pipeleers DG: Islet cell surface antibodies from insulin-dependent diabetics bind specifically to pancreatic B-cells: J. Clin. Invest. 70: 41-49, 1982.
64. Huen A, Haneda M, Freedman Z, Lernmark Å, Rubenstein AH: Quantitative determinations of islet cell surface antibodies using ^{125}I-protein A: Diabetes 32: 460-465, 1983.
65. Dyrberg T, Poussier P, Nakhooda F, Marliss EB, Lernmark Å: Islet cell surface and lymphocyte antibodies often precede the spontaneous diabetes in the BB rat: Diabetologia 26: 159-165, 1984.
66. Pipeleers D, Winkel van de M, Dyrberg T, Lernmark Å: The spontaneously diabetic BB rats have age-dependent B cell specific surface antibodies at the time of onset: (Submitted for publication), 1986.
67. Martin DR, Logothetopoulos J: Complement-fixing islet cell antibodies in the spontaneously diabetic BB rat: Diabetes 33: 93-96, 1984.
68. Dobersen MJ, Scharff JE, Ginsberg-Fellner FBS, Notkins AL: Cytotoxic autoantibodies to beta cells in the serum of patients with insulin-dependent diabetes mellitus: N. Engl. J. Med. 303: 1493-1498, 1980.
69. Hartling SG, Lindgren F, Dahlquist G, Thalme B, Möller E, Efendic S, Persson B, Binder C: Proinsulin - a possible subclinical indicator for B-cell dysfunction: Diabetologia 29: 202 A, 1986.
70. Lernmark Å, Kanatsuna T, Rubenstein AH, Steiner DF: Detection and possible functional influence of antibodies directed against pancreatic islet cell surface: Adv. Exp. Med. Biol. 119: 157-163, 1979.
71. Like AA, Appel MC, Rossini AA: Autoantibodies in the BB/W rat:

Diabetes 31: 313-318, 1982.
72. Elders M, MacLaren NK: Identification of profound peripheral T lymphocyte immunodeficiencies in the spontaneously diabetic BB rat: J. Immunol. 130: 1723-1731, 1983.
73. Contreas G, Lernmark Å, Mathiasen ER, Deckert T: Immune complexes in insulin-dependent diabetes: Biomed. Acta 44, 1: 129-132, 1985.
74. Boitard C, Debray-Sachs M, Pouplard A, Assan R, Hamburger J: Lymphocytes from diabetics suppress insulin release in vitro: Diabetologia 21: 41-46, 1981.
75. Sensi M, Pozzilli P, Gorsuch AN, Bottazzo GF, Cudworth AG: Increased killer cell activity in insulin-dependent (Type 1) diabetes mellitus: Diabetologia 20: 106-109, 1981.
76. Wilson RG, Anderson J, Shenton BK, White MD, Taylor RMR, Proud G: Natural killer cells in insulin-dependent diabetes mellitus: Br. Med. J. 293: 244, 1986.
77. MacKay P, Boulton A, Rabinovitch A: Lymphoid cells of BB/W diabetic rats are cytotoxic to islet beta cells in vitro: Diabetes 34: 706-709, 1985.
78. Koevary S, Rossini AA, Stollen W, Chick WL: Passive transfer of diabetes in the BB/W rat: Science 220: 727-728, 1983.
79. Handler ES, Mordes JP, Seals J, Koevary S, Like AA, Nakano K, Rossini AA: Diabetes in the Biobreeding/Worcester rat. Induction and acceleration by spleen cell-conditioned media: J. Clin. Invest. 76: 1692-1694, 1985.
80. Mordes JP, Handler ES, Like AA, Nakano K, Rossini AA: Irradiated lymphocytes do not adoptively transfer diabetes or prevent spontaneous disease in the BB/W rat: Metabolism 35: 552-554, 1986.
81. Feutren G, Assan R, Karsenty G, Du Rostu H, Sirmai J, Papoz L, Vialettes B, Vexiau P, Rodier M, Lallemand A, Bach J-F: Cyclosporin increases the rate and length of remissions in insulin-dependent diabetes of recent onset: Lancet ii: 119-123, 1986.
82. Jaworski MA, Honore L, Jewell LD, Metha JG, McGuire-Clark P, Schouls JJ, Yap WY: Cyclosporin prophylaxis induces long-term prevention of diabetes, and inhibits lymphocytic infiltration in multiple target tissues in the high-risk BB rat: Diabetes Research 3: 1-6, 1986.
83. Stiller CR, Laupacis A, Keown PA, Gardell C, Dupre J, Thibert P, Wall W: Cyclosporin: Action, pharmacokinetics and effect in the BB rat model: Metabolism 32 suppl. 1: 69-72, 1983.
84. Like AA, Kislauskis E, Williams RM, Rossini AA: Neonatal thymectomy prevents diabetes mellitus in the BB/W rat: Science 216: 644-646, 1982.
85. Naji A, Silvers WK, Bellgrau D, Anderson AO, Plotkin S, Barker CF: Prevention of diabetes in rats by bone marrow transplantation: Ann. Surg. 194: 328-338, 1981.
86. Brayman KL, Markmann J, Silvers WK, Barker CF, Naji A: Neonatal bone marrow transplantation and the prevention of diabetes in BB rats: Immunology of Diabetes, Edmonton, Canada: S39, 1986.
87. Rossini AA, Faustman D, Woda BA, Like AA, Szymanski I, Mordes J: Lymphocyte transfusions prevent diabetes in the Bio-Breeding/Worcester rat J. Clin. Invest. 74: 39-46, 1984.

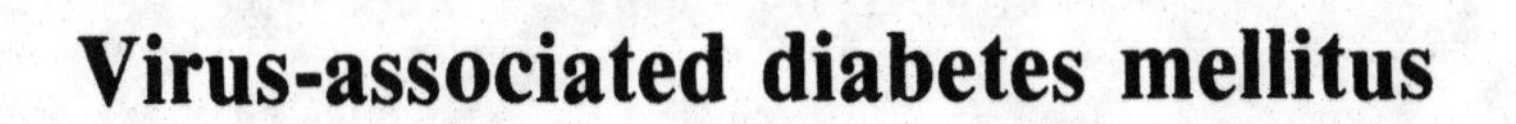

Virus-associated diabetes mellitus

7

ISLET CELL ANTIBODIES IN CHILDREN WITH MUMPS INFECTION

K. Helmke
Center of internal medicine, Justus-Liebig-University,
6300 Giessen, FRG.

ABSTRACT
Mumps infection was the first and still is the most frequently mentioned virus infection in connection with the development of typeI diabetes mellitus. In 21 mumps-infected children islet cell antibodies were demonstrated during 1979 to 1981. These positive results could not be obtained in the subsequent years. Islet cell surface antibodies, however were detectable in a high percentage of patients infected with mumps as well as with other viruses. All but one of those patients showing anti-islet cell immune activity did not develop diabetes mellitus. Their HLA constellation was not comparable with those of type I diabetic patients. In contrast the HLA constellation of 7 children who developed diabetes mellitus shortly after mumps vaccination did include DR4 in each case as well as DR3 in 3 cases.
These findings lead us to reconsider the role of viruses in inducing autoimmune reactions and possibly autoimmune diseases such as type I diabetes mellitus.

INTRODUCTION

A large number of recent investigations as well as a few earlier reports have shown autoimmune mechanism to play an important role in the pathogenesis of type I diabetes mellitus. Humoral and cellular islet and ß-cell specific immune reactions have been described in vitro as well as in vivo. Virus infections have repeatedly been suspected of trigger off the pathological autoimmune reactions.

Insulitis

The first indications pointing towards an inflammatory proc-

Becker, Y (ed), Virus Infections and Diabetes Melitus. ISBN 0-89838-970-4.

ess in the pathogenesis of diabetes mellitus date back to the beginning of the century. 1902 Schmidt (1) described lymphocytic infiltrates around the islets of Langerhans in juvenile diabetes mellitus.1940 von Mayenburg (2) described a similar observation and coined the term "insulitis". In 1965 Gepts (3) was able to confirm these observations by examining a number of juvenile diabetics having died soon after diagnosis of diabetes. Nearly 70% of these recent diabetics showed cellular infiltrates as described above. At a later stage of the disease these observations were made only rarely or not at all (4).

Immunephenomena

In vitro tests, too showed sensitisation of the immune system against islet cell antigens (migration inhibition, lymphocyte stimulation, T-cell cytotoxicity). The results were controversial, however (5-12). A change within the T-cell subpopulations - a decrease of suppressor cells and activation of helper cells - has been described as well as cytotoxic activity against human insulinoma cells in culture, and increased DR expression by the lymphocytes (13).

The discovery of islet cell antibodies put a new emphasis on humoral autoimmune factors in the etiology of type I diabetes mellitus (14,15). The cause of triggering factors of such an autoimmune process are at present merely subject to speculation (16).

Virus infection and diabetes mellitus

Reports describing a remarkable coincidence of a virus infection preceding the manifestation of diabetes mellitus date back to the previous century. The earliest reports in this respect concerned mumps infection: The Norwegian physician J. Stang in 1864 reported on a patient developing diabetes shortly after mumps infection. H. F. Harris in Philadelphia documented a similar case in 1899 (17). H. Kremer 1947, E. Henden 1962, W. M. McCrae 1963, Peig et al 1981 made similar observations (17-21). 1975 Sinaniotis et al reported the development of diabetes mellitus after mumps vaccination (22).

In recent years virus infections other than mumps such as coxsackie-B, rubella and german measles have been associated with diabetes manifestation as well (23-27).
Results of epidemiological studies have provided evidence for an association between mumps infection and diabetes manifestation. In 1927 - the preinsulin era - Gundersen reported an increased mortality rate caused by diabetes mellitus 2-4 years after a mumps epidemic (28). In 1975 H. A. Sultz and co-workers in Erie-County reported the incidence of diabetes to parallel that of mumps (29). Approximately 50% of 112 children with diabetes had had mumps or been exposed to mumps prior to diagnosis. The time interval between infection and manifestation of diabetes mellitus was about 3.8 years, however.

ISLET CELL ANTIBODIES (ICA) AND MUMPS

In 1979 we presented the case of a 5-year-old girl becoming diabetic 3 weeks after contracting mumps. The serum islet cell antibody (ICA) titre was high (30). Following on this observation we decided to examine the serum of patients being admitted to the hospital with mumps for islet cell antibodies in the years 1979 to 1984. This group of patients was rather a selected one as only patients with complications of mumps such as orchitis, meningitis or pancreatitis were admitted to the hospital.
We were able to demonstrate ICA in 21 out of 127 mumps-infected children. After 1981, however none of these children was positive for ICA anymore (Fig.1).
Also not a single remaining mumps-infected child developed diabetes mellitus. To the present day glucose metabolism in these children has been normal (31).

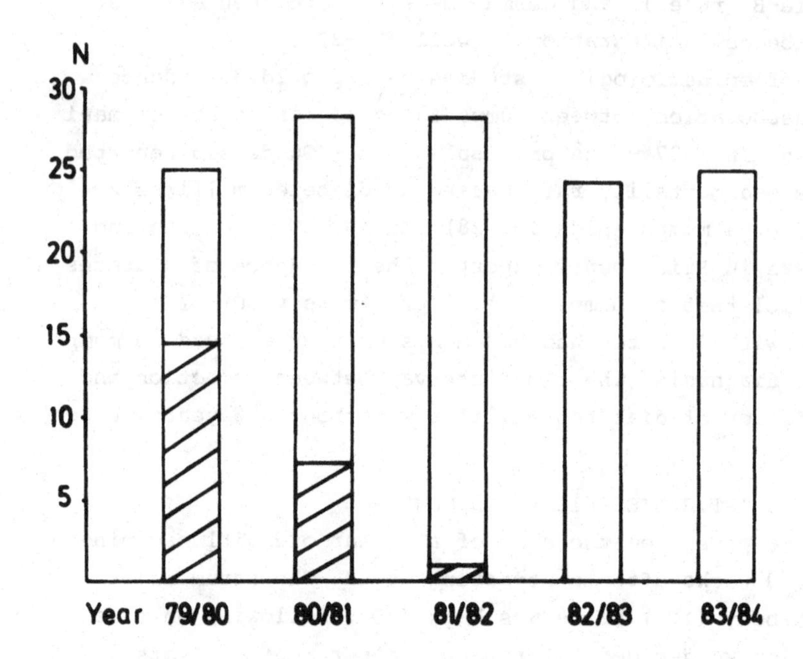

Fig. 1: Incidence of positive ICA finding in mumps-infected children during the years 1979 to 1981.
positive ▨ negative ☐

<u>Islet cell surface antibodies (ICSA) in virus infections</u>

In contrast with ICA the incidence of islet cell surface antibodies (ICSA) was relatively high in mumps as well as other virus-infected patients (31,32). 42 out of 68 mumps-infected children (62%), 32 out of 44 patients (73%) infected with enterovirus and 4 out of 11 with measles were positive for ICSA, although it was not possible to demonstrate ICA in sera of these patients. Patients suffering from type I diabetes mellitus and healthy controls showed ICSA in 70% and 14% respectively (Fig.2).

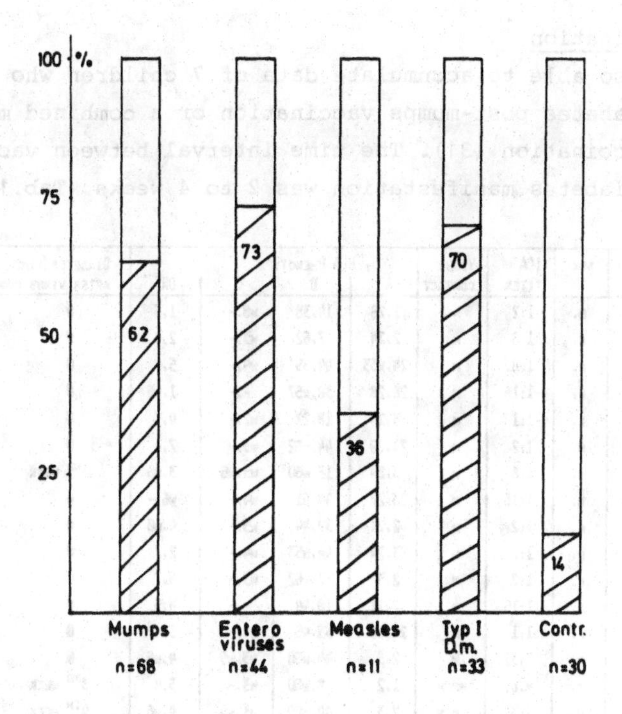

Fig. 2: ICSA in patients with viral infections, type I diabetes mellitus and controls.
positive [hatched box] negative [empty box]

These positive ICSA findings were neither restricted to a certain time limit as shown for ICA in mumps-infected children nor was there a direct correlation between virus titre and concentration of ICSA or ICA.
The follow-up of 15 mumps-infected patients over a period of 18 months uniformly showed a sharp decline in ICA titre over the first 3 months. A high persistant ICA titre over 9 months could only be demonstrated in the child with overt diabetes mellitus. ICSA showed no consistent behaviour over 18 months. A rise as well as a fall or even no change in concentration

was observed (32).

Mumps vaccination

We were also able to accumulate data of 7 children who developed diabetes post-mumps vaccination or a combined mumps-measles vaccination (31). The time interval between vaccination and diabetes manifestation was 2 to 4 weeks (Tab.1).

Patients	Age years	Sex	ICA titer	Virus contact	HLA Phenotype A	B	C	DR	Diabetes (time after virus cont.)
1	4	M.	1:2	M	1.29	14.38	w8.-	1.5	Ø
2	5	F.	1:8	M	2.24	7.62	w3.-	2.4	Ø
3	5	M.	1:8	M	24.w33	44.35	w4.-	5.-	Ø
4	5	F.	1:16	M	26.28	38.w57	-.-	1.w6	Ø
5	7	F.	1:1	M	3.24	18.27	w2.-	4.5	Ø
6	8	M.	1:2	M	24.29	44.w62	w3.-	7.-	Ø
7	8	M.	1:8	M	3.24	13.w60	w3.w6	3.w6	3rd week
8	9	M.	1:16	M	23.28	44.51	w4.-	w6.-	Ø
9	10	M.	1:2	M	2.24	39.44	w5.-	4.w8	Ø
10	10	F.	1:1	M	23.24	44.w53	w4.-	2.7	Ø
11	11	F.	1:2	M	2.3	7.w62	w3.-	5.7	Ø
12	12	M.	1:16	M	2.11	18.44	-.-	4.5	Ø
13	13	F.	1:1	M	24.30	35.45	w4.-	5.-	Ø
14	19	M.	1:32	M	2.3	44.w60	w3.w5	4.w6	Ø
15	2	F.	N.D.	MM V	1.2	8.w60	w3.-	3.4	3rd week
16	16	M.	1:32	M V	2.3	44.w60	w3.w5	4.w6	4th week
17	3	M.	1:2	MM V	2.32	8.w62	w3.-	3.4	2nd week
18	2	M.	1:8	MM V	2.24	44.51	-.-	4.w6	4th week
19	16	M.	N.D.	MM V	2.3	44.w60	w3.w7	4.w6	2nd week
20	2	F.	N.D.	M V	1.30	8.13	w4.w6	3.4	2nd week
21	2	F.	1:16	M V	2.3	w35.40	w3.w4	1.4	3rd week

Tab. 1: HLA distribution, ICA titre, age sex and virus contact in 14 mumps-infected and 7 mumps-vaccinated children. m= wild mumps infection, mv= mumps vaccination, mmv= mumps-measles vaccination

Each case examined was positive for ICA for a short period of time. Prospective studies were not possible here since observations were only made after diabetes manifestation.

HLA-frequencies

Non diabetic ICA-positive mumps infected children did not

show an increased frequency of HLA-DR3 or DR4 as often found in type I diabetes mellitus. In direct contrast to this all the children who developed diabetes mellitus after mumps vaccination and the child becoming diabetic after mumps infection were DR4 positive, 3 were shown to have DR3 as well and 6 had Cw3 in addition (Tab.1).

Islet cell toxicity in vitro

The role of humoral immune phenomena in the development of diabetes mellitus is not entirely clear. Although the majority of reports favour a minor role only, certain in vitro investigations have shown islet cell function and viability to be influenced by ICA or ICSA positive sera (34-39). This could be indicative of humoral immune phenomena contributing to the persistence of the disease.

On testing the sera of mumps- and enterovirus-infected patients for islet-cell toxicity, the results were rather surprising in that they were comparable with those of diabetics (Fig.3).

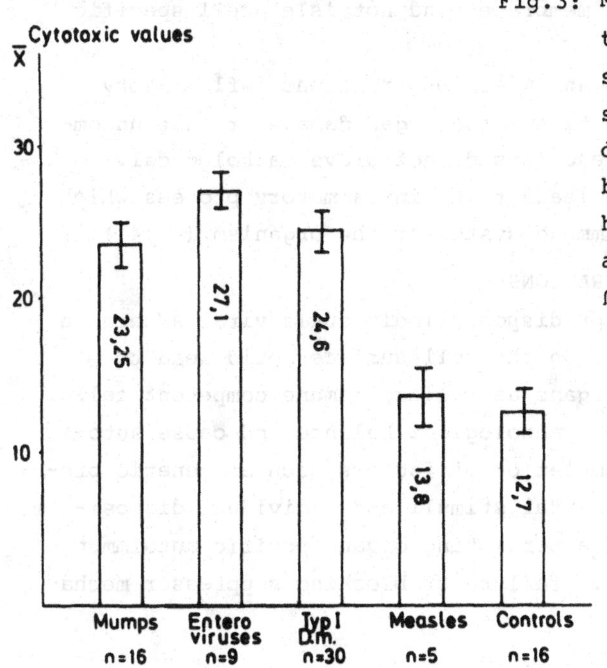

Fig.3: Mean values of cytotoxic reactions of sera from patients suffering from viral disease, type I diabetes mellitus and healthy controls against isolated ß-cells.

Mumps and enterovirus-infected patients showed islet-cell cytotoxicity in a number of patients. Sera of a small group of measles infected patients, however did not show more cytotoxicity than did the control group (32).

Reaction against common cell surface antigen

The finding of positive ICSA in non-diabetic virus-infected patients is as unexpected as the islet-cell cytotoxicity described above. There are some facts however,pointing towards these phenomena to be cell non-specific reactions with surface antigens induced by binding of viruses to the cell surface. Absorption studies which showed a marked reduction of antibody titre post-absorption with different cell types support the above hypothesis. It is probable for identical antigens to be found on the cell surface - the cell membrane - of different cell types. These could be demonstrated employing monoclonal antibodies, too. Therefore, overlap reactions are conceivable (40,41). The apparent antigen-specific immune reaction in vitro could thus be viewed as a reaction against "common cell membrane" and not islet-cell specific antigens.

During the course of an infection transient inflammatory processes as well as tissue and organ damage are not uncommon. Usually these reactions do not prove pathological. The damage to tissue leads to an inflammatory process which gets halted by the immune system of the organism (40,41).

PATHOGENETIC CONSIDERATIONS

In some genetically predisposed individuals viral adherance to HLA constellations on the cell surface could lead to damage to specific organs as well as immune competent cells. This may disrupt the immunological balance and cause autoaggression. The accumulation of factors such as genetic predisposition, environmental stimuli and individual disposition may trigger off a persisting organ-specific autoimmune process e.g. selective failure of blocking suppressor mechanisms (42,43).

Mumps and diabetes

The finding of ICA in mumps-infected children is surprising for two reasons: Firstly because it seems to be a temporary phenomenon, and secondly as only one child developed diabetes. Various case reports stating an association between mumps infection and the development of diabetes are conducive to speculating that there may be a causal connection with the induction of a process of autoaggression. On the other hand, mumps infection is very common and its coincidence with another common disease such as diabetes mellitus may not be of significance. In order to answer this question a prospective study of considerable dimensions is required. To our knowledge such a study has not been performed.

Vaandrager et al in 1986 did a prospective study on a smaller collective of patients (44). 242 sera of 184 mumps-infected patients were investigated one to six weeks after becoming symptomatic. ICA were not found, 2 children showed antibodies against glucagon cells. 149 children aged 16 to 21 months did not have ICA before or 8 weeks and 2.5 years after mumps vaccination.None of these children developed diabetes mellitus. Similar findings were reported by other groups (45). In contrast to this finding Ratzmann et al (46) reported a high incidence of ICA after mumps infection. None of these patients became diabetic either. Fixed pancreatic sections were used by this group to demonstrate ICA. However, this procedure is not suitable for ICA demonstration purposes in our opinion because of a high degree of non-specificity which has been documented by other investigators as well (47-49). Therefore, the results are not comparable with those of the former authors.

All in all manifestation of diabetes mellitus as well as temporary ICA positivity after mumps infection and mumps vaccination has only been observed sporadically. The typical HLA-constellation was found only in the overtly diabetic children, not in the mumps-infected ICA-positive children. These observations and facts taken together indicate that additional predisposing factors are required in the development of diabetes mellitus. These may be additional environmental influences apart from genetic and individual organ predisposition (50). The environmental stimuli responsible for the induction of mechanisms of autoaggression remain to be elucidated. In any case the agents blamed are much more common than diabetes. The possibility of a synergistic interplay with other factors in genetically predisposed individuals can, however not be excluded (50).

Autoaggression, a new concept?

In 1984 G.F. Bottazzo and B. M. Dean were able to demonstrate the induction of DR-expression on endocrine thyroid and later islets of Langerhans ß-cells (50,51). They believed this finding to be the key to the development of organspecific autoaggression (52). On the basis of this hypothesis one could postulate the following: Certain viruses are able to penetrate the endocrine tissue of genetically predisposed individuals without necessarily causing symptoms or signs of disease. This leads to an increased gamma-interferon production which in turn causes HLA-DR expression on the cell surface. These extraordinary DR molecules allow presentation of cell surface antigens to autoreactive T-cells which secrete more interferon. In this way the DR antigen expression is sustained and further stimulation and activation of B and T cells is warranted (53). Perpetuation of this process is only conceivable in the presence of defective suppressor activity of the immune system i.e. immunological imbalance. The defect has to be selective and may possibly be of importance regarding organspecificity because of other mechanisms not being affected (42,43, 54-56).

We do not at present have a founded explanation for the temporary occurrence of ICA in mumps-infected children. Different species of mumps viruses present at different points in time as shown by monoclonal antibodies may be of importance (57). In our study a few of many thousands of children vaccinated against mumps developed diabetes and only one child became diabetic after mumps infection. In spite of this it is remarkable to note the association repeatedly reported of mumps and lately also coxsackie-B virus infection with diabetes mellitus (26).

Virus induced autoimmunity

Virus infections were reported shortly i.e. 2-4 weeks before diabetes manifestation. During the last couple of years, however evidence of longterm autoimmune destruction of islet cells prior to diabetes manifestation has accumulated (25, 50,58). Diabetes only becomes manifest after about 95% of destruction of islet-cell tissue has occured. It is unlikely, therefore for the reported cases of diabetes manifestation following mumps vaccination or infection to be linked directly. As far as the autoimmune process is concerned this may be the last straw. It could be due to stimulation of the immune system during an infection leading to specific interaction.

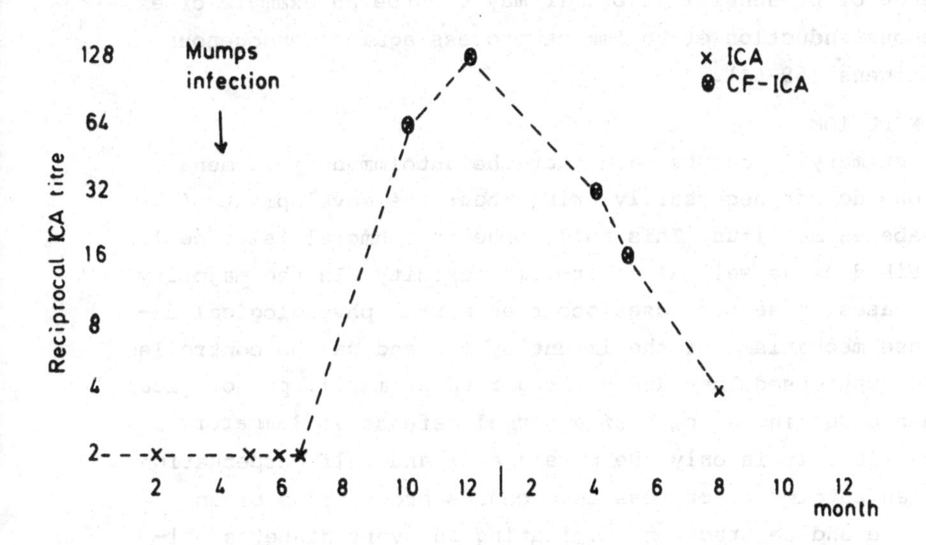

Fig. 4: ICA in a ten-year-old diabetic boy after mumps infection.

An isolated observation made during our mumps study may be interpreted as an example of such a specific interaction between exogenous mumps infection and production of organ-specific ICA (Fig.4).

A 10-year-old boy with diabetes of 8 years duration went down with mumps. Investigations for islet-cell antibodies over a time span of 2 years before the mumps infection produced borderline to negative results. 6 weeks after mumps infection ICA were markedly positive and complement fixing. ICA remained positive for 6 months and the titre declined after this. Complement fixation, too could no longer be demonstrated by simple immunofluorescence. Three months later the state of borderline to negative ICA concentration was reached again and persisted. No further signs of polyclonal immune activation was shown. These observations could possibly be interpreted to be an indication for the interaction of certain viral species with lymphocytic cell clones in the presence of presensitisation. It may thus be an example of exogenous induction of an immune process against endogenous antigens (59-61).

CONCLUSION

In summary it can be said that the autoimmune phenomena alone do not necessarily bring about the development of diabetes mellitus. This holds true for humoral islet-cell antibodies as well as islet-cell toxicity. In the majority of cases these processes occur as normal physiological defense mechanisms of the immune system and can be controlled and suppressed. They are not regarded primarily pathological when occurring as part of a normal defense inflammatory reaction. It is only the persistence and self perpetuation of an autoimmune process that causes progressive organ damage and destruction culminating in overt diabetes mellitus. A selective defect of the suppressor activity as described for other autoimmune diseases may be the cause.
A possible viral induction of this process may have occured years before, maybe even in utero.

In conclusion therefore, the autoimmune phenomena following virus infection and those present at the time of diabetes mellitus manifestation could provide a clue to the mechanisms of induction of the disease rather than being directly pathogenic themselves. Our observation as well as those described in the literature may be interpreted in this way.

Acknowledgment

I am grateful to I. Schneider-Waterberg for helpful discussion and correction of the manuscript and to S. Scherer for skillful technical assistence and careful preparation of manuscript and figures.

1) Schmidt, M.B.: Münch. med. Wschr. 49, 51, 1902

2) von Meyenburg H.: Schweiz. med. Wschr. 70,554, 1940

3) Gepts, W.: Diabetes 14, 619 - 633, 1965

4) Gepts, W., H. De J. Mey, Diabetes 27, 251-262, 1978

5) Nerup, J., Andersen,O., Bendixen, G., Egeberg, J. Poulsen, J.E., Diabetes 20, 424-427, 1971

6) Mascart-Lemone F., Delespesse G., Dorchy H., Lemiere B and Servais,G. Clin. Exp. Immunol. 47, 296-300, 1982

7) Galluzo, A., Giordano, C., Rubino, G., Bompiani, GD: Diabetologia, 26, 426-430, 1984

8) Pozzilli, P., Zuccarino, O., Iavioli, M., Andreani, D., Sensi, M., Spencer, KM., Bottazzo, GF, Beverly, PCL, Kyner JL, Cudworth AG: Diabetes, 32, 91-94, 1983

9) Herold, K., Huen,A., Gould, L., Traisman, H., Rubenstein AH: Diabetologia 27, Suppl., 102-105, 1984

10) Ilonen, J., Surcel HM, Mustonen, A., Kaar, M.L, Akerblom, H.K.: Diabetologia, 27, Suppl, 106-108, 1984

11) Gupta, S., Fikrig S.M., Khanna, S., Orti, E: Immunol. Lett., 4, 289-294, 1982

12) Pozzilli, P., Sensi, M., Dean, B., Gorsuch AN, Cudworth AG: Lancet, 2, 173-175, 1979

13) Jackson, RA, Morris, MA., Haynes BF, Eisenbarth, GS: N. Engl. J. Med. 306, 785-788, 1982

14) Bottazzo, G., F., A., Florin-Christensen, D.Doniach: Lancet 2, 1279-1282, 1974

15) Bottazzo, G.F., R. Pujol-Borell, D. Doniach: Clin. Immunol. Allergy, 1, 139 - 142, 1981

16) Champsaur, H.F., G.F. Bottazzo: J. of Pediatrics, 100, 15-20, 1982

17) Harris, H.F.: Boston med. Surg. J. 140, 465-469, 1899

18) Kremer, H.V.,: Am. J. Med. 3, 257-258, 1947

19) Henden, E.: Lancet 1, 1381, 1962

20) McCrae, W.M.: Lancet 1300-1301, 1963

21) Peig, M., Ercilla, G., Milian, M., Gomis, R.: Lancet II, 1007, 1981

22) Sinaniotis, C.A., Daskalopoulon, E., Lapatsanis, P., Doxiadis, S.,: Arch. Dis. Chilhood 50, 749-750, 1975

23) Toniolo, A., Onodera, T.:In: Immunology in Diabetes Kimpton Medical Publications, London, 1978, pp. 71-93

24) Menser, M.A., Forrest, J.M., Bransby, R.D.: Lancet I, 57-60, 1978

25) Gamble, D.R.,: Brit. Med. J. 2, 99-101, 1980

26) Rayfield, E.J., Seto, Y.,: Diabetes 27, 1126-1137,1978

27) Federlin, K., Helmke, K.(Eds.) Behring Inst. Res., Commun. 75, 1984

28) Gundersen, E.: J. Infect. Dis. 41, 197-202, 1927

29) Sultz, H.A., Hart, B.A., Zielezny, M & Schlesinger, E.R.: J. Ped. 86, 654-656, 1975

30) Helmke, K., Otten, A., Willems, W.: Lancet II, 211-212, 1980

31) Helmke, K., Otten, A., Willems, W.R., Brockhaus, R., Müller-Eckhardt, G., Stief, T., Federlin, K: Diabetologia 29, 30-33, 1986

32) Helmke, K., Weimer, R., Willems, W.W., Otten, A., Mäser, E., Velcovsky, H.G., Federlin, K.: Deutsche Med. Wschr. 10, 369-373, 1986

33) Müller-Eckhardt, G., Stief, T., Otten, A., Helmke, K. Willems, W.R., Müller-Eckhardt, C.: Immunobiol. 167, 338-344, 1984

34) Boitard, C., Sai, P., Debray-Sachs, Assan, R.,: Clin. Exp. Immunol., 55, 571-580, 1984

35) Dobersen, M.J. & Scharff, J.E.: Diabetes 31, 459-462, 1982

36) Kanatsuna, T., Lernmark, A., Rubenstein, A.H. & Steiner: Diabetes 30, 231-235, 1981

37) Rittenhouse, H.G., Oxender, D.L., Pek, S & Ar, D.: Diabetes 29, 317-320, 1980

38) Kanatsuna, T., Baekkeskov, S., Lernmark, A & Ludvigsson, J.: Diabetes 32, 520-524, 1983

39) Helmke, K., Seitz, M., Brockhaus, R., Weimer, R. Otten, A & Federlin, K.: Immun. Infekt. 11, 199-208, 1983

40) Lamb JR, Eckles DD, Lake D., Johnson AH, Hartzman RJ, Weedy JN: J. Immunol. 128, 233-235, 1982

41) Cooke, A., Lydyard, PM, Roitt, IM: Immunol. Today, 4, 170-175, 1983

42) Buschard, K., Madsbad, S., Rygaard, J.: J. Clin. Lab. Immunol. 8, 19-29, 1983

43) Lernmark, A., In: Immunology and Diabetes. (Eds. Andreani, D., di Mario, U., Federlin, K. and Heding, L.H.), Kimpton, London 1984, pp 121-131

44) Vaandrager, G.J., Molenaar, J.L., Bruining, G.J., Plantinga, A.D. and Ruitenberg, E.J., Diabetologia 29, 406, 1986 (letter)

45) Richens, E.R. & Jones, W.G.: Lancet I, 507-508, 1981

46) Ratzmann, K.P., Strese, J., Witt, S., Berling, H., Keilacker, H., Michaelis, D.: Diabetes Care 7, 170-173, 1984

47) Dobersen, M.J. Bell, M., Jenson, AB, Notkins AL, Ginsberg-Fellner, F.: Lancet 2, 1078, 1979(letter)

48) Dean, B., Pujol-Borrell, R., Bottazzo, GF: Lancet, 2, 1343-1344, 1982, (letter)

49) Rosenbloom, AL: Lancet, 1, 72-73, 1983, (letter)

50) Bottazzo, G.F., Pujol-Borrell R. and Gale, E., In: The Diabetes Annual (Eds: Alberti, K.G., M.M. and Krall, L.P.) Elsevier Science Publishers, B.V. Amsterdam, 1985 pp16-52

51) Bottazzo, GF., Dean BM: Diabetologia, 27, 259 A,1984

52) Bottazzo, GF. Todd, I., Pujol-Borrell, R.: Immunol. Today, 5, 230 - 231, 1984

53) Zinkernagel, RM., Doherty, PC: Nature (London), 251, 547-549, 1974

54) Horita, M., Suzuki, H., Onodera, T., Ginsberg-Fellner F., Fauci, AS, Notkins, AL: J. Immunol., 129, 1426-1429, 1982

55) Fairchild, RS., Kyner, JL, Abdon, NI: J. Lab. Clin. Med. 99, 175-184, 1982

56) Topliss, D., How, T., Lewis, M., Row, V., Volpe, R.: J. Clin. Endocrinol. Metabol. 57, 700-705, 1983

57) Orvell, C.: J. Immunol. 132, 2622-2624 , 1984

58) Gorsuch, A.N., Soencer, K.M., Lister, J., McNally, J.M., Dean, B.M., Bottazzo, G.F., Cudworth, A.G.: Lancet II, 1363-1365, 1981

59) Helmke, K., Federlin, K., Otten, A. and Willems, W.R.: Diabetologia 29, 407, 1986 (letter)

60) Plotz, PH.: Lancet, 2, 824-826, 1983

61) Adams, DD., Adams YJ. Knight, JG., McCall, J., White, P., HorrocKs, R., Loghem, E.: Lancet, 1, 420-423, 1984

8

HERPES VIRUS INFECTIONS IN HUMAN PANCREAS ISLETS AND ANIMAL MODELS

T. KURATA[1], T. SATA[1], T. IWASAKI[2], T. ONODERA[3] and A. TONIOLO[4]

Department of Pathology, National Institute of Health, Kamiosaki 2-10-35, Shinagawaku, Tokyo 141, Japan[1]; Department of Pathology, Iwate Medical University, Uchimaru 19-1, Morioka 020, Japan[2]; Laboratory of Immunology, National Institute of Animal Health, Jyosuihoncho 1500, Kodaira, Tokyo 187, Japan[3]; Institute of Microbiology, University of Pisa Medical School, 13, Via S. Zeno, 56100 Pisa, Italy[4].

ABSTRACT

The pancreatic tissue from 131 human cases with fatal herpes infections were viropathologically examined for islet lesions. Diagnoses of all cases were confirmed by virus isolation, serological tests and immunohistochemistry. Focal or massive necrosis, including destruction of beta cells, intranuclear inclusion bodies and inflammatory changes in islets were found in 45 cases, in accordance with the existence of viral antigens. All pancreatic lesions were observed as the expression of a generalized herpesvirus infection and not as solitary damage appearing only in the pancreas. The findings suggest the participation of these viruses in damage to beta cells resulting in diabetes.

If the animal is first treated with a subdiabetogenic dose of beta cell toxin or reovirus, and then injected with murine cytomegalovirus that normally produces little or no diabetes, the cumulative insults induce diabetes. The degu, a rat-like hystricomorph, showed spontaneous diabetes mellitus with characteristic cytomegalovirus inclusion bodies in islet cells.

Becker, Y (ed), Virus Infections and Diabetes Melitus. ISBN 0-89838-970-4.

INTRODUCTION

The development of juvenile or other insulin-dependent diabetes in man has been suspected of having an infectious etiology, namely mumps, coxsackie, rubella or other viruses (1). A link between virus infection and the onset of diabetes is supported by a reduction in the number of beta cells, as well as by inflammatory changes that occur in Langerhans islets (2, 3). Pancreatic lesions in herpesvirus infections (herpes simplex virus: HSV, varicella zoster virus: VZV; and cytomegalovirus: CMV) have frequently been found at autopsy of cases infected *in utero* or at the perinatal stage, or after intensive treatment with immunosuppressant or antineoplastic drugs (4). Vertical transmission of CMV infection during pregnancy causes congenital cytomegalic inclusion disease, and perinatal infection produces latency of the virus in some undetermined organs. HSV produces generalized neonatal infection, encephalitis, genital herpes, mucocutaneous and other lesions, and VZV causes varicella in children and herpes zoster in adults. These two viruses become latent in the sensory ganglia after primary or secondary infection. In immunocompromised patients, reactivation of these latent infections, or reinfection from outside, frequently results in fatal dissemination of the virus. In spite of several clinical and pathological studies, no direct viropathological evidence has been presented for the viral etiology of diabetes mellitus, since biopsy of the pancreas is not usually done.

In mice, several viruses can cause diabetes (5). Encephalomyocarditis (EMC) virus infects pancareatic beta cells and induces a severe form of diabetes in certain inbred strains. Characteristic inclusions have also been found in beta cells of islets in immunosuppressed mice infected with murine CMV (MCMV) (6). In some circumstances, reoviruses also infect beta cells and trigger an autoimmune response resulting in a mild form of diabetes, characterized by abnormalities in the glucose tolerance test that generally disappear within several weeks (7). Diabetes may also be triggered through cumulative environmental insults (8, 9). In the animal model, the severity of diabetes depends upon the degree of beta cell damage.

In this chapter, we describe islet lesions in the human pancreas caused by herpesvirus infection and animal models of MCMV-induced diabetes mellitus.

HUMAN CASES AND METHODS

Human cases

The clinical and pathological records in our laboratory were reviewed and 131 cases were chosen for examination, out of approximately 1500 cases that were sent from the hospitals and institutions in metropolitan and other areas of Japan, to determine the virus infection viropathologically. All selected cases had generalized herpesvirus infections in which two or more organs were involved, and viral infections were confirmed at least by viral antigens and intranuclear inclusion bodies. Cases with inadequate slides or blocks of the pancreas were excluded. The number of cases examined is listed in Table 1. No solitary lesion was recognized in the pancreas. Several hundred more cases were also excluded, because viral antigens were not demonstrated or the pancreas sections were not available.

Immunohistochemistry and histopathology

Formalin-fixed and paraffin-embedded materials from all organs including the pancreas of 131 cases were examined for viral antigens of herpes group viruses. Diagnoses were determined by virus isolation, serological tests, and immunofluorescence, in cases where frozen materials were available. The methods of antigen detection were previously described in detail (10, 11). The indirect immunofluorescence and immunoperoxidase method (ABC) were applied for paraffin sections after trypsin treatment to enhance antigenicity (11). All sections were stained with all four antisera to HSV (type 1 and 2), VZV and CMV. Recent advances in immunohistochemistry have made it possible to survey restrospectively those cases in which viropathological studies were not performed before or during autopsy (10, 12). Serial sections were stained with hematoxylin and eosin (HE) for histopathological study.

RESULTS

The examined cases with fatal generalized herpesvirus infection are summarized in Table 1. In all cases, herpesvirus infections were determined by immunohistochemistry and other virological methods. Forty-five specimens of pancreas had viral antigens in the islets with or without characteristic histological findings. In 86 pancreatic sections, antigens and other histological changes were not recognized. Cases of dual infection were not included. The distribution of viral antigens of HSV or

Table 1. Human cases with herpes virus infections confirmed by immunohistochemistry and histopathology

Virus*	Total cases	Islet lesion
H S V infection	32	6
V Z V	31	21
C M V	58	13
Dual infection		
C M V + H S V	8	3 (C M V, H S V)
C M V + V Z V	1	1 (V Z V)
C M V + Adeno virus	1	1 (C M V)

* Viral antigens were found in several organs by immunofluorescence and/or immunoperoxidase method (ABC). Pancreatic islet lesions were confirmed by direct comparison of histopathological findings with existence of viral antigens.

VZV in internal organs differ from each other. The former invades paranchymal cells more frequently, and the latter the interstitial tissues. It is quite difficult to differentiate between epidermal lesions by histopathology, since intranuclear inclusion bodies of HSV closely resemble those of VZV. These two viral infections should thus be identified by the presence of viral antigens in tissues. CMV infection can be diagnosed without difficulty because of pathognomonic cytomegalic cells with large inclusions.

HSV infection

Thirty-two cases with generalized HSV infection were submitted for immunohistochemical study, and HSV antigen was detected in the islets of six cases with or without histological changes (Table 2). Five were neonates less than 11 d of age and all isolated viruses were type 1. Seven cases with type 2 infection showed no antigen in islets. Immunofluorescence and/or the ABC method revealed HSV antigen in the islets, ducts, acini, interstitium and vascular endothelia. Diffuse massive or focal necroses with hemorrhages, eosinophilic degeneration and interstitial edema were recognized histopathologically. Inflammatory cell reaction

Table 2. Herpes simplex virus infection with pancreas islet lesion

Case	Type isolated	Age	Sex
1.	1	11 d	F
2.	1	8 d	F
3.	1	11 d	F
4.	1	5 d	F
5.	1	5 d	F
6.	n.d.*	19 yr	M (De Sanctis-Cacchione syndrome)

*n.d.: not done

was not seen. Cowdry type A inclusion bodies in the nuclei were clearly observed in islet cells in Case 1, but giant cell formation was not found. The other four cases had no inclusion bodies in the islets, but these were clearly present in other areas and other tissues. Usually, the main lesions in generalized neonatal HSV infections are hepatoadrenal necroses which are surrounded by hemorrhages and characteristic intranuclear inclusion-bearing cells. Viral antigens and histological changes were also observed in the pituitary glands, brain, lungs, spleen, trachea, kidneys, esophagus, thymus, thyroid gland and intestines (Figs. 1-5).

In case 6, the patient with De Sanctis-Cacchione syndrome died of generalized candidiasis with HSV infection in the tongue, tonsils and pancreas. Interstitial hemorrhages in the pancreas were prominent, and antigen was detected, in spite of the absence of nuclear inclusion bodies. Neither diabetes symptoms nor dermal lesions were recognized clinically in any of the cases examined.

VZV infection

Pancreatic sections were examined in 31 cases with disseminated VZV infection confirmed by both immunohistochemistry and other methods. Islet lesions with viral antigens were revealed in 21 cases (Table 3). Fourteen cases were associated with malignancies, leukemia or lymphoma, and two nephrotic syndrome. Only one case (No. 17) had no underlying disease. Diabetes was not noted in any cases. All cases had skin eruptions prior to VZV dissemination, but this was not the case for HSV infection in which dermal lesions were not always present. All patients died within 3-14 days after the appearance of eruptions. Varicella-type infection or reactivation of latent VZV in the sensory ganglia with its dissemination in the body occurs in immunocompromised individuals. The clinical records

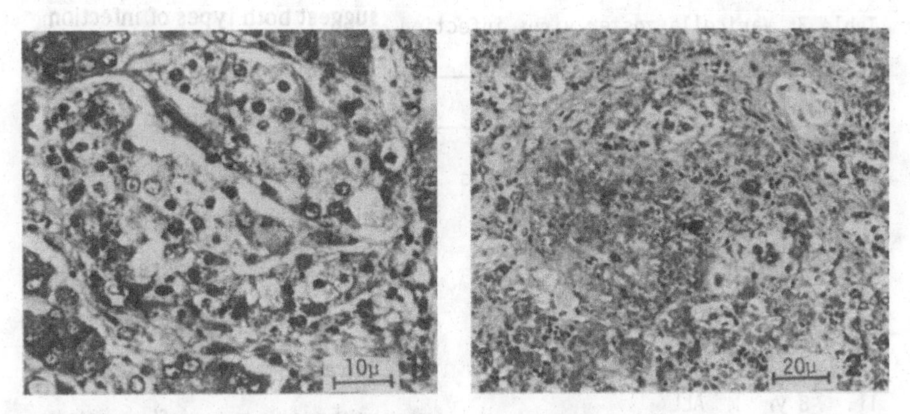

Fig. 1. Pancreatic islet in HSV infection. Several intranuclear inclusions are seen in islet cells. No necrotic changes are observed. HE.
Fig. 2. Most of an islet becomes necrotic with residual cells bearing intranuclear inclusions in HSV infection. HE.

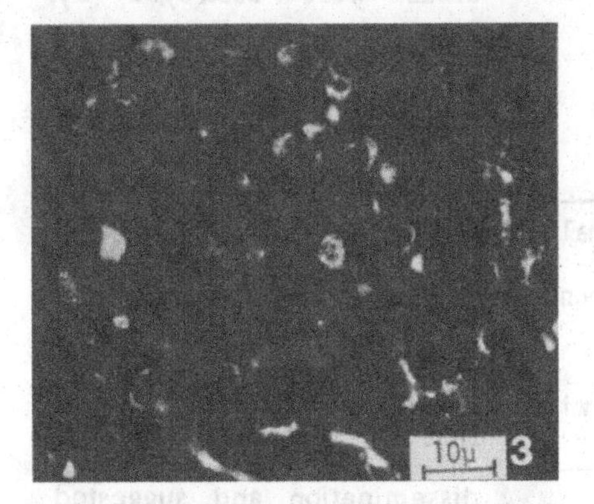

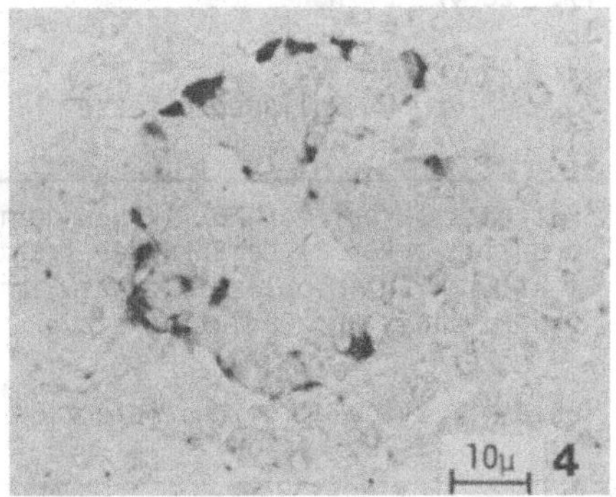

Fig. 3. HSV antigen is positive in the intranuclear inclusions and cytoplasm of the islet cells. Immunofluorescence.
Fig. 4. HSV antigen was only positive in the cytoplasm of the peripheral islet cells. Immunoperoxidase stain.

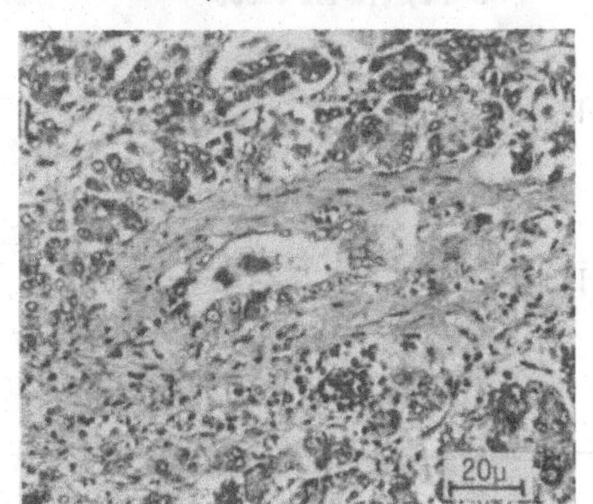

Fig. 5. Endothelial linings have intranuclear inclusions in the small muscular artery. HE.

Table 3. Varicella zoster virus infection and pancreas islet lesions

Case	Age	Sex	Underlying disease*
1.	31 yr	F	T-cell lymphoma
2.	48 yr	M	Diffuse bronchiolitis
3.	34 yr	F	Malignant lymphoma
4.	35 yr	M	Hodgkin's disease
5.	75 yr	F	Multiple myeloma
6.	4 yr	M	Nephrosis
7.	40 yr	M	AML + BMT
8.	15 yr	M	Nephrotic syndrome
9.	4 yr	F	ALL
10.	11 yr	M	Medulloblastoma
11.	8 yr	M	ALL
12.	5 yr	F	ALL
13.	64 yr	M	Renal cell carcinoma
14.	6 yr	F	AML
15.	7 yr	M	ITP
16.	22 yr	F	Malignant lymphoma
17.	26 yr	F	None
18.	6 yr	F	ALL
19.	5 yr	M	ALL
20.	47 yr	F	Collagen disease
21.	4 yr	F	AML

*Patients except for No. 17 had dermal eruption prior to death under treatment with antineoplastic drugs, immunosuppressants and radiation etc.

Table 4. Cytomegalovirus infection with islet lesion

Case	Age	Sex	Underlying disease
1.	5 mo	F	Congenital CID*, hepatosplenomegaly
2.	45 d	M	Congenital CID, hepatosplenomegaly
3.	50 d	F	Congenital CID, premature baby
4.	81 yr	M	CMV disease of the colon
5.	61 yr	M	Chronic cholecystitis
6.	57 yr	M	Lung cancer
7.	66 yr	F	Still's disease
8.	64 yr	M	Thymic tumor
9.	51 yr	F	Malignant plexus papilloma
10.	65 yr	M	Hypoglycemia
11.	70 yr	M	AML
12.	28 yr	M	AIDS
13.	62 yr	F	Tbc meningitis

*CID: Cytomegalic inclusion disease

suggest both types of infection. Distribution of viral antigens in islets, interstitium, acini or ducts was compatible with the histopathological changes, including degeneration, coagulation necrosis and intranuclear inclusion bodies without inflammatory reaction. In some islets, entire areas were destroyed by hemorrhages surrounded by cell debris. Hyalinization was found in islets of case 13. Endothelia of small-sized muscular arteries had inclusion bodies with antigen that resembled HSV dissemination and suggested generalized viremia (Figs. 6-11). In six cases (Nos. 1, 6, 8, 11, 17 and 20), viral antigens were detected not only in the pancreas, but also in the liver, stomach, kidneys, intestines, rectum, adrenals, salivery glands, bone marrow, testes, urinary

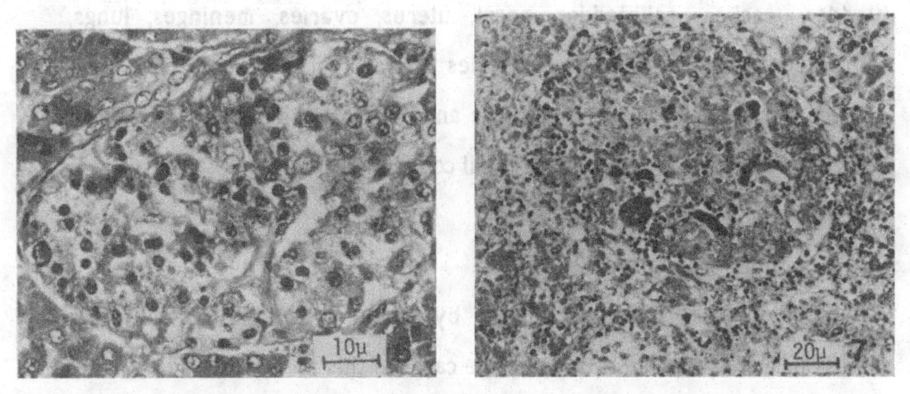

Fig. 6. VZV infection in islet. Intranuclear inclusions were observed in some islet cells without necrotic changes. HE.
Fig. 7. VZV infection causes necrotic changes with hemorrhage in an islet surrounded by cell debris. HE.

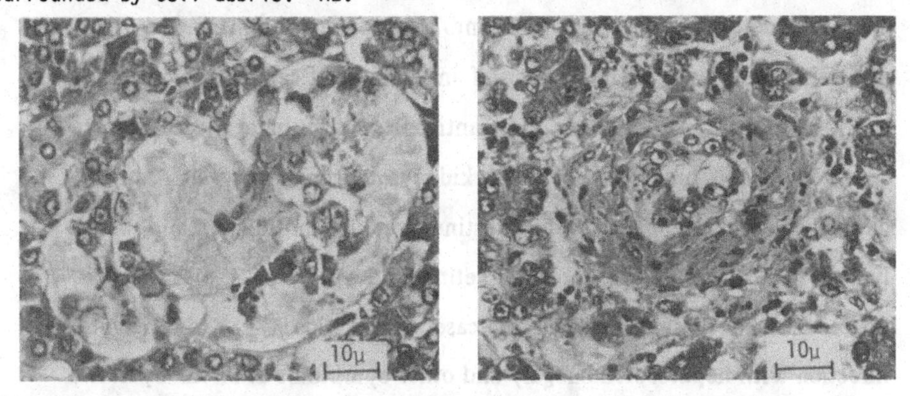

Fig. 8. Hyalinosis is observed in an islet in a case of VZV infection. HE.
Fig. 9. Endothelial involvement is seen in a small muscular artery in VZV infection. HE.

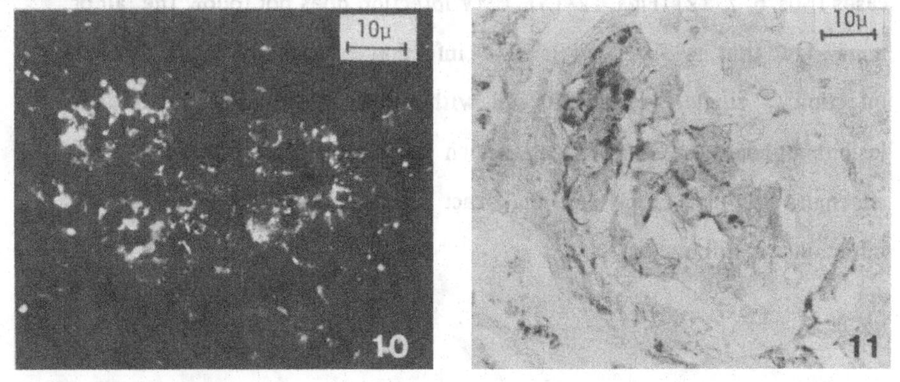

Fig. 10. VZV antigen is observed in the nuclei and cytoplasm of the islet cells. Immunofluorescence.
Fig. 11. Immunoperoxidase stain shows VZV antigen in the cytoplasm of the islet cells.

bladder, trachea, gallbladder, heart, uterus, ovaries, meninges, lungs, thyroid, pituitary gland, lymph nodes and dorsal root and trigeminal ganglia. In tonsils, tongue, esophagus and skin, infection was observed in squamous epithelium and subepithelial connective tissues.

CMV infection

CMV infection is characterized by giant cells with intranuclear inclusions and cytoplasmic bodies. The cases in this study were selected by this histopathological marker, as well as by immunohistochemistry. In some islets, the viral antigen was detected without clear cytomegalic cells. We could detect the viral antigens and/or cytomegalic cells in the islets, acini and ducts without infiltration of inflammatory cells in 13 out of 58 cases (Table 4). In addition, viral antigens and cytomegalic cells were revealed in the lungs, liver, heart, kidneys, adrenals, thyroid, salivary glands, spleen, tongue, stomach, intestines, rectum, pituitary gland, brain, lymph nodes, thymus, esophagus, retina and Auerbach's plexus. The degree of change varied from case to case. Three (Nos. 1-3) had congenital infection with hepatosplenomegaly and other symptoms at birth. Systemic severe CMV infection was revealed in these congenital and several adult cases (Nos. 6, 7, 12) (Figs. 12-15). CMV infection does not follow the "all or none" law that is observed in VZV infection. In many cases of CMV infection, a single cytomegalic cell without the associated histological lesions appears in some organs, such as lungs, gastrointestinal tract, adrenals, etc. The latent site and exact reactivation mechanisms of CMV infection are still unclear.

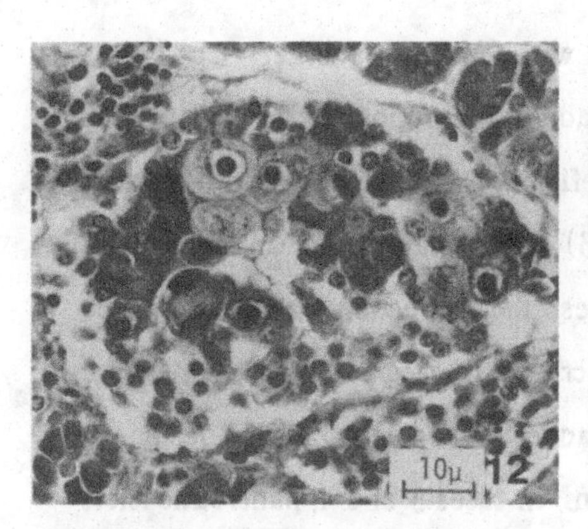

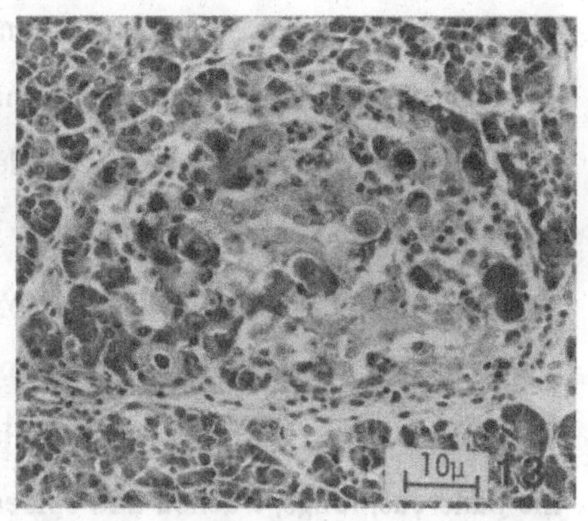

Fig. 12. Islet change in CMV infection. Several cytomegalic cells with Cowdry A intranuclear inclusions are observed in an islet without necrotic change. HE.
Fig. 13. Focal necrosis is observed in a lower part of the islet with several cytomegalic cells. HE.

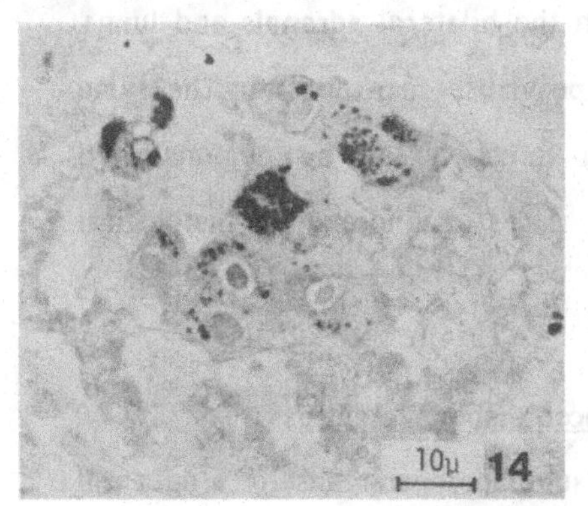

Fig. 14. CMV antigen is observed as granules in the cytoplasm of cytomegalic cells. Immunoperoxidase stain.

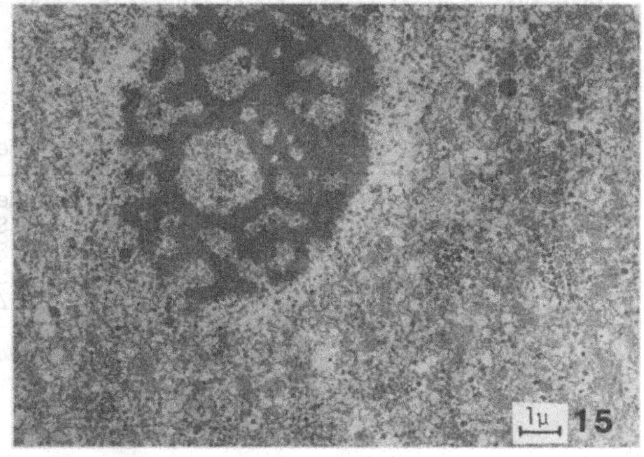

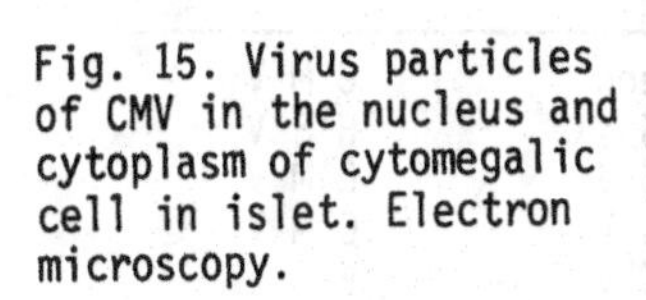

Fig. 15. Virus particles of CMV in the nucleus and cytoplasm of cytomegalic cell in islet. Electron microscopy.

Dual infection

We had ten cases of dual infection with CMV and another virus (Table 5). Diagnosis of dual infection was made by immunohistochemistry. Case 1, with severe combined immunodeficiency, received a bone marrow transplantation. Adenovirus (type 3) was isolated and antigens were observed in the lungs, trachea and eyes; CMV infection was shown in the lungs, spleen, salivary glands and pancreas. In case 2, CMV lesions were seen in the lungs, adrenals, urinary bladder and spleen, and HSV lesions in the pancreas, lungs, trachea and spleen. In cases 3 and 4, pancreatic islets were damaged by CMV. In case 5, VZV affected the islets with diffuse massive necrosis and hemorrhages, and also the tonsils, left adrenal gland and tongue. CMV also made lesions in the bilateral adrenals and lungs. Combined infection with different herpesviruses can occur in the same organ or in different organs. Dual infection can be demonstrated immunohistochemically in misleading cases diagnosed by histological examination only.

Animal models in cytomegalovirus-induced diabetes mellitus

If an animal is first treated with a subdiabetogenic dose of a beta cell toxin such as streptozotocin, which reduces the beta cell reserve, and then

Table 5. Dual infection by cytomegalovirus and another virus

Case	Age	Sex	Underlying disease	Dual infection*	Islet
1.	8 mo	M	S C I D + B M T	C M V + Adeno	C M V
2.	26 yr	M	A L L + B M T	C M V + H S V	H S V
3.	79 yr	F	Malignant lymphoma	C M V + H S V	C M V
4.	63 yr	F	Agranulocytosis	C M V + H S V	C M V
5.	43 yr	F	Adult T-cell leukemia	C M V+ V Z V	V Z V

*Viral antigens of one or two viruses were detected at the same time in one or more organs.

infected with viruses that normally produce little or no diabetes, the cumulative insults induce diabetes. This section shows that murine cytomegalovirus (MCMV) infection leads to mild focal beta cell damage, and that multiple environmental insults increase the severity of diabetes. The degu (*Octodon degus*), a rat-like hystricomorph, showed spontaneous diabetes mellitus with characteristic cytomegalovirus inclusion bodies in islet cells.

MCMV and cumulative environmental insults

Newborn NIH Swiss male and female mice wereinfected with the Smith strain of MCMV (9) or Lang strain of reovirus type 1 that had been passaged more than seven times in mouse pancreatic beta cell cultues to increase their tropism for beta cells. Two or four weeks later, the mice were challenged with either a subdiabaetogenic dose of streptozotocin (1 mg/mouse), which did not alter glucose homeostasis in preliminary experiments, or with the D variant of EMC virus (5), or MCMV. Each animal was then bled four times; glucose from non-fasting animals was measured after 7 and 14 days, and single-point 60-min glucose tolerance tests were performed 10 and 17 days after infection. Uninfected animals and animals given only streptozotocin were bled at the same time. The results were expressed as the glucose index (13). An animal was considered diabetic if its glucose index was equal to, or exceeded, 3 S.D. above the mean of uninfected mice.

Table 6 shows that when NIH Swiss male mice were infected with only MCMV or reovirus, or given only streptozotocin, diabetes did not develop. In contrast, when males were infected first with MCMV or reovirus and then challenged with streptozotocin, 84 and 93%, respectively, of animals developed diabetes. When males were first infected with MCMV or

Table 6. Glucose abnormalities produced by multiple enviromental insults in male mice.

First infection		Challenge				
CMV	Reovirus	Strepto-zotocin	EMC virus	CMV	Glucose index	Percent** diabetic
+	-	-	-	-	153± 16*	0
-	+	-	-	-	137± 19	0
-	-	+	-	-	178± 10	0
-	-	-	+	-	256± 81	56
+	-	+	-	-	313± 88	84
+	-	-	+	-	373± 70	100
-	+	+	-	-	255± 34	93
-	+	-	+	-	446±112	100
-	+	-	-	+	185± 29	5

NIH Swiss mice were infected with $2x10^5$ PFU of MCMV or reovirus, and 4 weeks later were challenged with sub-diabetogenic dose of streptozotocin or $1x10^4$ PFU of EMC virus, or with $2x10^5$ PFU of MCMV 2 weeks later. Approximately 20 mice were tested in each group. *Mean ± S.D. **Percentage of mice with glucose index 3 S.D. above the mean of uninfected controls i.e. males 164± 17, females 141± 13.

Table 7. Glucose abnormalities produced by multiple enviromental insults in female mice.

First infection		Challenge				
CMV	Reovirus	Strepto-zotocin	EMC virus	CMV	Glucose index	percent** diabetic
+	-	-	-	-	124± 34	0
-	+	-	-	-	133± 19	0
-	-	+	-	-	152± 12	3
-	-	-	+	-	182± 83	20
+	-	+	-	-	253± 45	93
+	-	-	+	-	214± 43	81
-	+	+	-	-	204± 54	80
-	+	-	+	-	303±121	96
-	+	-	-	+	164± 26	22

See the foot note of Table 6.

reovirus and then challenged with EMC virus, the percentage of animals developing diabetes increased from 56% with EMC virus alone to 100%. However, mice first infected with reovirus and then challenged with MCMV showed a relatively small increase in the glucose index as compared with controls. The reverse experiment was not done, as older mice are less susceptible to reovirus infection (14). The trend in female mice was similar to that in male mice, except that the glucose indices were generally lower (5) (Table 7).

To determine if streptozotocin enhanced virus-induced diabetes by increasing the susceptibility of islet cells, e.g. by immunosuppressing the host (15), the number of EMC-infected cells were determined by staining sections of the pancreas with fluorescein-labeled anti-EMC antibody (9). Table 8 shows that approximately the same number of, or slightly fewer, islet cells were infected in streptozotocin-treated mice as in untreated mice. This suggests that streptozotocin acts by decreasing the beta cell reserve rather than by increasing the susceptibility of beta cells to infection. Microscopic examination of sections of pancreas from mice treated with streptozotocin showed that some beta cells had been destroyed. Furthermore, quantification by radioimmunoassay revealed that the insulin content of the pancreas of mice treated with 1 or 2 mg of streptozotocin was reduced by 15 and 48%, respectively, 12 days later.

It is known that in mice MCMV infects pancreatic acinar cells, but there is little information about the effect of the virus on islets (6). In man, inclusion bodies are sometimes found in the islets of Langerhans of patients suffering from overwhelming CMV infection (2). In the present study, histopathological examination of islets from MCMV-infected mice revealed focal cellular infiltrates and pyknotic nuclei in a small percentage (i.e. 5-15%) of the islets at 7-14 days after infection. Staining of the islets with a fluorescein-labeled anti-MCMV antibody revealed viral antigen in 10% or less of the islet cells. The mild and focal islet cell damage produced by MCMV does not appear to be sufficient to leave the mice diabetic at the end of one month, unless the animals are challenged with virus or chemicals that add to the degree of beta cell damage. The fact that neither MCMV nor reovirus produced severe beta cell damage may explain why only a few mice develop diabetes when first infected with reovirus and then challenged with MCMV.

Depletion of beta cells in mice by a subdiabetogenic dose of streptozotocin may provide a useful model for identifying other diabetogenic viruses, especially those which infect and destroy only a minimal number of beta cells. The enhancement of virus-induced diabetes by low doses of streptozotocin also raises the possibility that, in humans, a series of viral infections or other environmental insults, e.g. chemicals, drugs or toxins, each producing some beta cell damage, finally results in overt insulin-dependent diabetes, once the beta cell reserve has been sufficiently depleted.

Table 8. Percentage of cells in islets of Langerhans containing viral antigen.

Mouse strain	Pretreated with streptozotocin	% infected islet cells		
		days after infection		
		2	3	4
C57BL/6	+	2	12	7
C57BL/6	-	2	12	9
SJL/J	+	30	36	10
SJL/J	-	41	44	12

Each mouse received sub-diabetogenic dose of streptozotocin intraperitoneally 12 days before infection with 1×10^4 PFU of EMC virus. At 2, 3 and 4 days after infection, sections of pancreas from groups of three mice were stained with fluorescein-labeled anti-EMC antibody and approximate number of cells containing viral antigen were determined. At each time point, an average of 40 islets and over 2,000 cells were counted (Reprinted with permission from ref. 13).

Cytomegalovirus-induced diabetes in *Octodon degus*

Spontaneous diabetes mellitus was occasionally observed in the rat-like degu (16). Characteristic CMV inclusion bodies were seen in islet, acinar and ductal cells of the pancreas (17). The pancreas in 14 out of 281 animals examined contained CMV inclusion bodies. These 14 degus and ten of the others also had inclusions in the salivary glands. Ten of out of these 24 degus had clinical signs of diabetes. The electron microscopic study demonstrated many spherical DNA virus-like particles, 200 nanometers in size in pancreatic tissues from a diabetic degu.

Clinical evidence of diabetes in the degu was indicated by glucosuria, elevated fasting blood glucose, or abnormal glucose tolerance in 41 out of 281 (15%) degus. CMV inclusions were seen in the pancreas of seven degus with and without clinical evidence of diabetes.

Hyperplasia and hypertrophy of islets occurred in the 14 degus with inclusions in the pancreas. Amyloidosis of the islets was noted in six of 24 degus examined, four of which had inclusions in the pancreatic tissue.

The insulitis in degus with CMV infections is morphologically similar to that in juvenile-onset diabetes in man (3). The presence of insulitis, characteristic inclusions of CMV in the pancreas, and clinical evidence of islet dysfunction suggest cytomegalovirus as the cause of diabetes mellitus in the degu. Usually, human neonates and immunodeficient patients with generalized CMV infections succumb; thus, evaluation of future islet cell function in man is not possible. The effect of CMV infection on islet cell function in man is not fully understood, and the degu could serve as an animal model for the study of a possible relationship between CMV and diabetes.

ACKNOWLEDGMENTS

We are grateful to Emeritus Professor Dr. Yuzo Aoyama of the University of Tokyo, and Dr. Abner Louis Notkins, Scientific Director of the National Institute of Dental Research, National Institutes of Health, Bethesda, Maryland, for their helpful comments and advice.

REFERENCES

1. Notkins, A.L. Sci. Am. 241:62-73, 1979.
2. Jenson, A.B., Rosenberg, H.S. and Notkins, A.L. Lancet 2:354-358, 1980.
3. Gepts, W. Diabetes 14:619-633, 1965.
4. Wong, D.T. and Ogra, P.L. Clin. Med. North Am. 67:1075-1092, 1983.
5. Yoon, J.W., McClintock, P.R., Onodera, T. and Notkins, A.L. J. Exp. Med. 152:878-892, 1980.
6. Craighead, J.E. In: The Diabetic Pancreas (Eds. Volk B.W. and Wellman K.F.), Plenum Press, New York, 1977, pp. 467-488.
7. Onodera, T., Ray, U.R., Meletz, K.A., Suzuki, H., Toniolo, A. and Notkins, A.L. Nature 297:66-68, 1982.
8. Toniolo, A., Onodera, T., Yoon, J.W. and Notkins, A.L. Nature 288:383-385, 1980.
9. Onodera, T., Suzuki, H., Toniolo, A. and Notkins, A.L. Diabetologia 24:219, 1983.
10. Kurata, T., Hondo, R., Sato, S., Oda, A., Aoyama, Y. and McCormic, J.B. Ann. N.Y. Acad. Sci. 420:192-207, 1983.
11. Sata, T., Kurata, T., Aoyama, Y., Sakaguchi, M., Yamanouchi, K. and Takeda, K. Virchows Arch. (Pathol. Anat.) 1986, in press.
12. Aoyama, Y. and Kurata, T. In: Immunofluorescence in Medical Science (Eds. Kawamura, A. and Aoyama, Y.) University of Tokyo Press, pp.101-158.
13. Ross, M.E., Onodera, T., Brown, K.S. and Notkins, A.L. Diabetes 25:290-197, 1976.
14. Stanley, N.F. Br. Med. Bull 23:150-155, 1967.
15. Saiki, O., Negoro, S., Tsuyuguchi, I. and Yamamura, Y. Infect. Immun. 28:127-131, 1980.
16. Liss, R.H., Fox, J.G., Charest, M. and Murphy, J.C. J. Cell Biol. 79:404, 1978.
17. Fox, J.G. and Murphy, J.C. Vet. Pathol. 16:625-628, 1979.

9

PANCREATIC CELL DAMAGE IN CHILDREN WITH FATAL VIRAL INFECTIONS

B. JENSON AND H. ROSENBERG

Department of Pathology, Georgetown University Schools of Medicine and Dentistry, Washington, D.C., and Department of Pathology and Laboratory Medicine, University of Texas Medical School at Houston, Houston, Texas

ABSTRACT

The rare fatalities from viral infections most commonly occur in neonates and immunosuppressed older children and adults. With a fatal viremia, pathologic changes in pancreatic acini, interstitium, and islets are frequently subtle and may be overlooked. Islet cell degeneration and necrosis with or without associated inflammation is usually associated with viral infections or ingestion of chemicals that are cytotoxic for islets. Insulitis without degeneration or necrosis of islet cells can probably be attributed to an autoimmune phenomenon. Coxsackievirus B, cytomegalovirus, congenital rubella, and varicella-zoster were the four viruses most frequently associated with islet cell cytopathology. Two of these, Coxsackievirus B and rubella, have been implicated as a trigger for development of IDDM. Insulin degranulation of beta cells in Coxsackievirus infections was greater than expected from the degree of cytopathology in the islets; this was not the case with the other fatal viral infections. Conversely, 50% or fewer beta cells were degranulated in one case each of cytomegalovirus and Coxsackievirus B_4 infections associated with acute-onset IDDM; two patients with congenital rubella syndrome, one with acute-onset IDDM and another with long-standing IDDM had no insulin-producing beta cells. These findings suggest that in some cases of IDDM associated with acute viral infections, insulin-producing beta cells may be initially preserved and potentially amenable to protection by therapeutic measures.

INTRODUCTION

The infrequent fatalities in humans due to viral infections usually occur in the neonate or immunosuppressed older children and adults. Clinically, except for mumps, involvement of the pancreas as part of a generalized viral infection is rarely appreciated. Morphologically, as judged from the reports of fatal cases, the pattern of pancreatic

Becker, Y (ed), Virus Infections and Diabetes Melitus. ISBN 0-89838-970-4.

involvement includes inflammation and necrosis of the acini, interstitium, and particularly the islets (1-16). In a review of about 10,000 autopsy records on infants and children from several children's hospitals and referral centers (1), 150 cases (1.5%) had well-documented disseminated viral infections due to 14 different viruses as determined from the clinical records, virus isolation studies, serological titers, and histopathologic features of viral infections. An additional 100 cases were included in the survey but will not be included in this chapter because the virus was not identified and there was no evidence of pancreatic viral cytopathology.

Of the 150 children with fatal viral infections, 28 had cytopathology of the islets of Langerhans attributed to four different viruses: 1) 4 of 7 cases of Coxsackievirus B (Cx B) infection; 2) 20 of 45 cases of cytomegalovirus (CMV) infection; 3) 2 of 14 cases of varicella-zoster (V-Z) infection, and 2 of 45 cases of congenital rubella syndrome (CRS) (Table 1). One of children with CMV infection and an additional case with CRS (7) died within a week after acute onset of IDDM; the child with CMV had prominent inclusions in the islets (Fig. 1) and the child with CRS had chronic insulitis. Of the 150 fatal viral infections only 3 cases, each with V-Z, produced inflammation of the acini and interstitium without insulitis. This concordance between insular and acinar inflammation conforms with the derivation of the islets during embryogenesis from totipotential cells in duct and periacinar cells and the potential for sharing some of the same viral receptors which are exposed during viremia. Only 20% (31 of 150) of fatal cases had morphologic evidence of insulitis and/or pancreatitis.

Table 1. Cytopathology of islets in fatal viral infections

	Cytopathology	Total
Coxsackievirus B	4	7
Cytomegalovirus	20	45
Varicella-Zoster	2	14
Rubella	2	45

Viral infections have been proposed as the trigger for most cases of IDDM based on epidemiologic and serologic data, clinical case reports, *in vitro* viral infections of human pancreatic beta cells in tissue culture, and the murine models for virus-induced diabetes. Some epidemics of mumps, rubella and infectious hepatitis have been followed by an increased incidence of IDDM but it is not clear whether the epidemics directed attention towards associated cases of IDDM or whether the viruses resonsible for the epidemics were diabetogenic (reviewed in ref. 3). Few morphologic studies have been reported on patients dying of concurrent acute-onset IDDM and documented viral insulitis and pancreatitis. In the animal model, development of a diabetes-like state appears to depend on the extent of the initial infection and associated destruction of insulin-producing beta cells, whereas in most humans the putative initial infection by any of multiple viruses only triggers a common pathway of autoimmune beta cell destruction leading to IDDM (3).

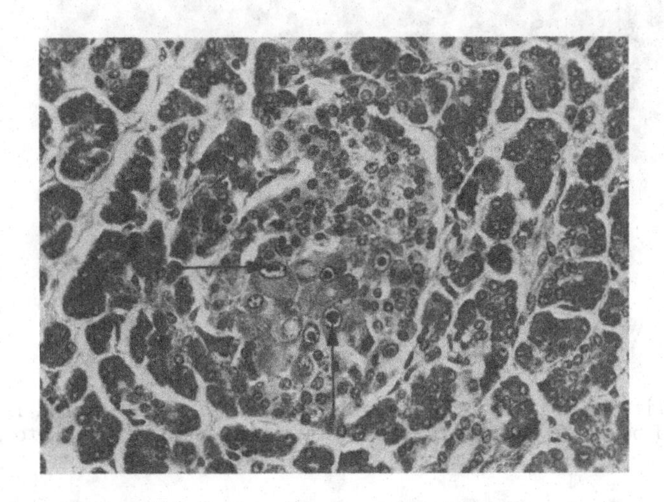

Fig. 1. Typical CMV inclusions (arrows) without associated insulitis in islet of 11-year-old male with disseminated CMV infection and diabetic ketoacidosis after treatment with steroids (Reprinted from ref. 1).

MORPHOLOGY OF VIRAL PANCREATITIS

For a variety of reasons, pancreatic disease may be overlooked during pathologic studies of disseminated viral infections. Even with proper tissue preservation, the pathologic changes are often subtle, focal, and with a minimal or absent cellular inflammatory response, perhaps because most fatalities occur in patients who are neonates or immunosuppressed. Even with an intact inflammatory response, infection and necrosis of islet cells may not elicit an insulitis. Pancreatic islets, separated from the interstitium by a network of capillaries, may have a rapid and effective mechanism for removal of degenerating or dead cells before an autoimmune response ensues.

For example, with encephalomyocarditis-induced murine islet damage and diabetes mellitus in otherwise healthy animals, macrophages rapidly remove necrotic cells without evidence of inflammation (17). Insulitis characterized by chronic inflammation but without necrosis of islet cells has a presumed autoimmune basis and almost invariably relates to acute-onset IDDM (18, 19). Insulitis characterized by degeneration and necrosis and an inflammatory cell infiltrate suggests the onset of IDDM during a systemic viral infection (Fig. 2) or damage by a beta cell toxin (Fig. 3), e.g., Vacor, a commercially available rat poison that kills by causing diabetes (1-3).

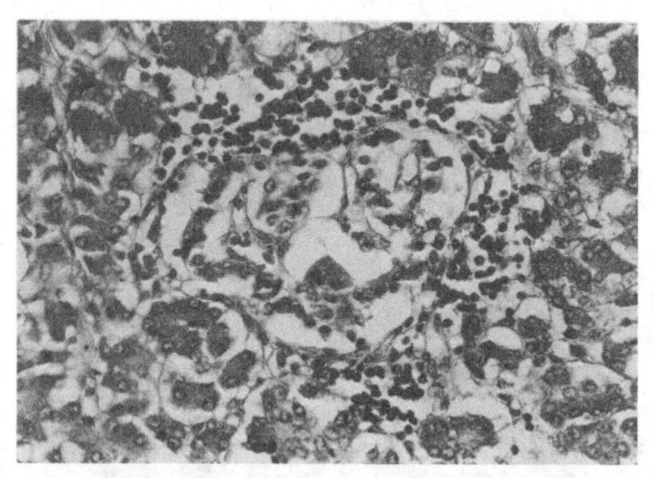

Fig. 2. Insulitis consisting of mantle of small lymphocytes around islet of 10-year-old boy with Coxsackievirus B_4 infection and diabetic keto-acidosis. (Reprinted by permission from ref. 2).

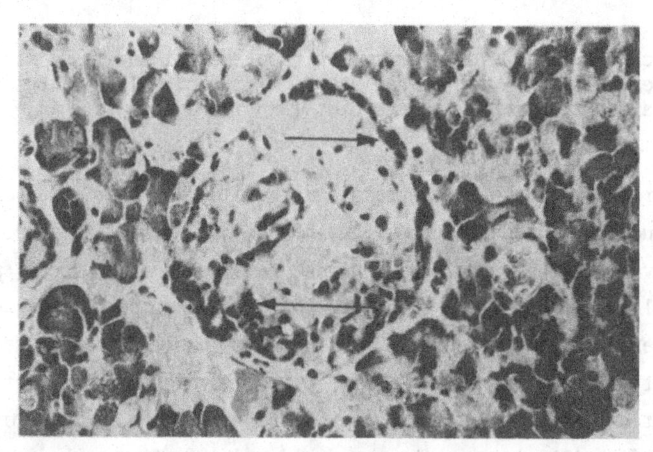

Fig. 3. Coagulation necrosis of beta cells in center of islet leaving a rim of intact alpha cells (arrows) following accidental ingestion of Vacor by young child.

Generally, viruses affect the pancreas in a manner similar to other organs of the same patient. Viruses such as the hepatitis viruses and rubella may not elicit specific cytopathic changes but may initiate an associated autoimmune response (Fig. 4). The herpes viruses, particularly CMV and V-Z cause distinctive pathological findings. CMV infection is characterized by markedly enlarged cells, large pleomorphic nuclei containing basophilic inclusions surrounded by a clear halo, and smaller cytoplasmic inclusions (Figs. 5 and 6). V-Z infections are characterized by normal-sized cells with intranuclear but no cytoplasmic inclusions, and various stages of associated degeneration and necrosis (Fig. 7); multinucleated giant cells characteristic of V-Z infection in the skin did not appear in the pancreas. Cx. B group infections produce a spectrum of lesions varying from minimal degeneration of islet cells without inflammation to mild or moderate islet cell necrosis and associated infiltration of lymphocytes, macrophages and neutrophils (Figs. 8-11).

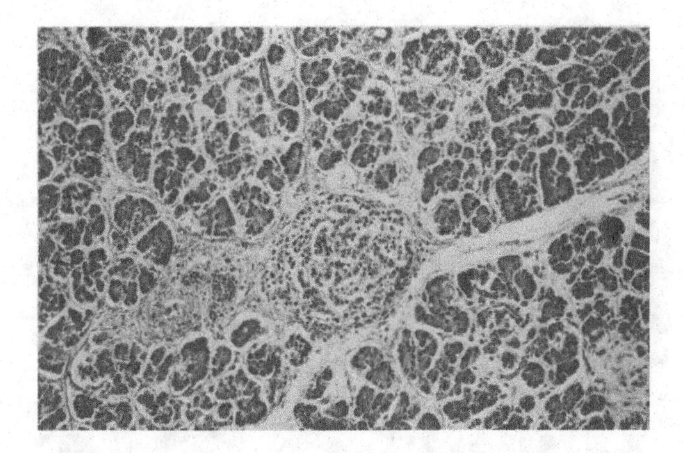

Fig. 4. Insulitis in 6-week-old girl with congenital rubella. Cords of islet cells are separated from one another by edema and an infiltrate of mixed inflammatory cells.

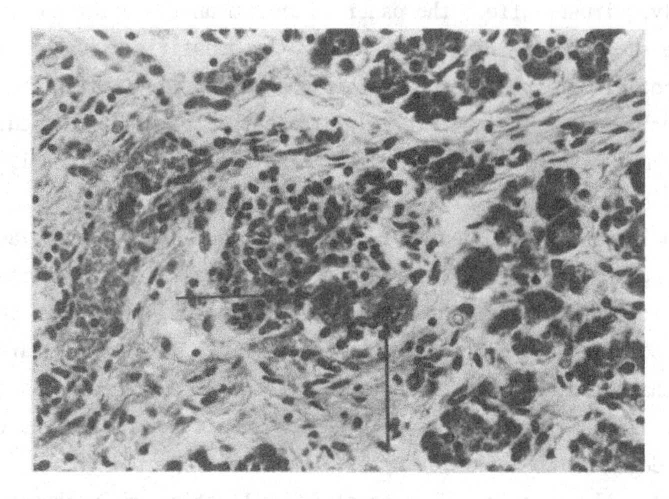

Fig. 5. Intranuclear CMV inclusion bodies (arrows) are seen in an islet of a neonate with congenital CMV infection.

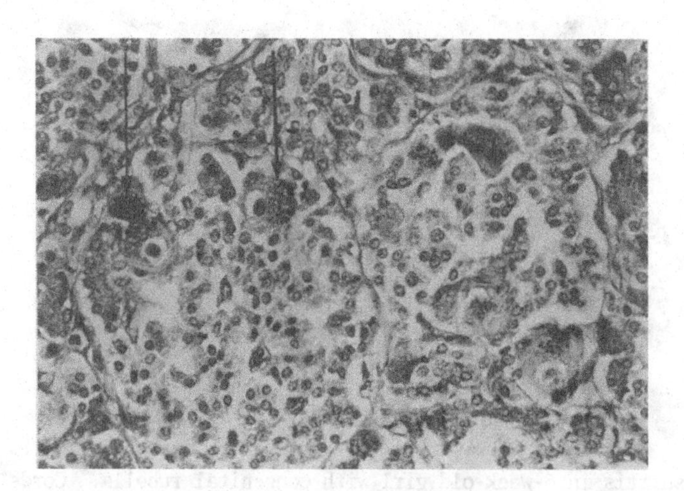

Fig. 6. Intranuclear inclusion bodies with associated cytoplasmic inclusions (arrows) are seen in islet of neonate with disseminated CMV infection. (PAS stain).

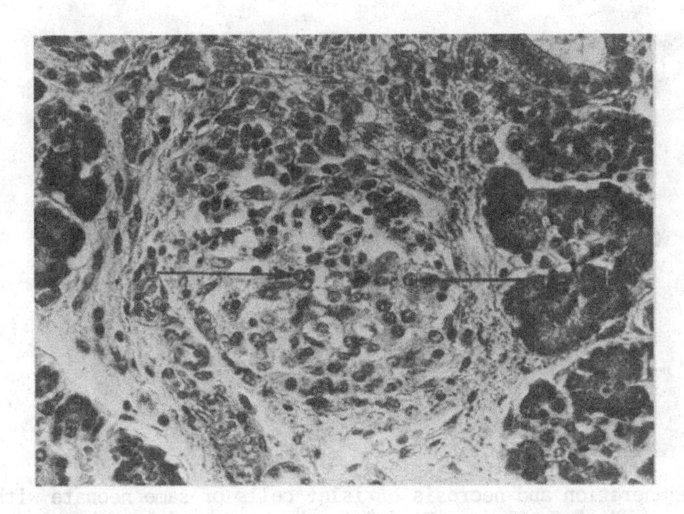

Fig. 7. Congenital varicella-zoster infection in a neonate. The upper third of islet appears normal, the middle third has typical varicella-zoster intranuclear inclusions (arrows) and the lower third contains cells with pyknotic nuclei and scanty cytoplasm. (Reprinted from ref. 1).

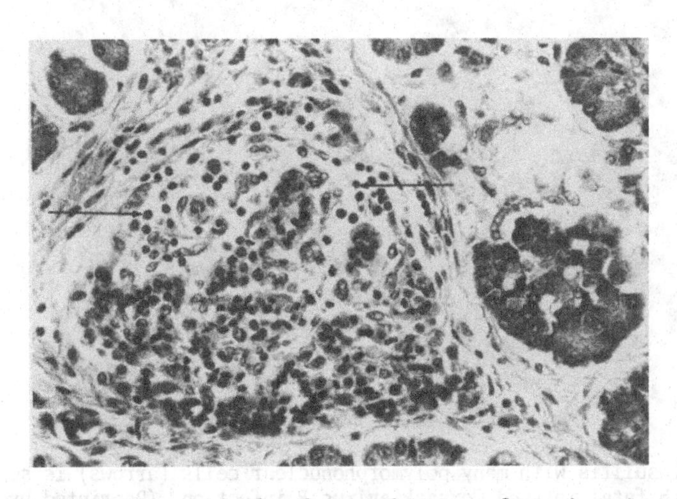

Fig. 8. Insulitis composed of a few lymphocytes of varying sizes (arrows) without associated necrosis are seen in islet of neonate with Coxsackievirus B_4 infection. (Reprinted by permission from ref. 2).

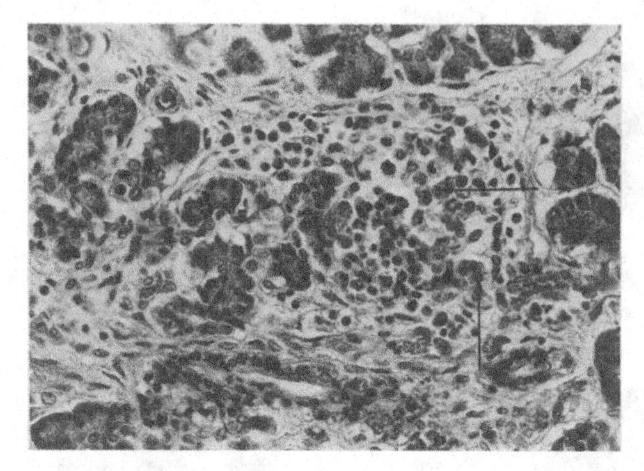

Fig. 9, Degeneration and necrosis of islet cells of same neonate with Coxsackievirus B_4 infection as Fig. 8 is characterized by pyknotic nuclei of varying size (arrows) and condensed eosinophilic cytoplasm. (Reprinted by permission from ref. 2).

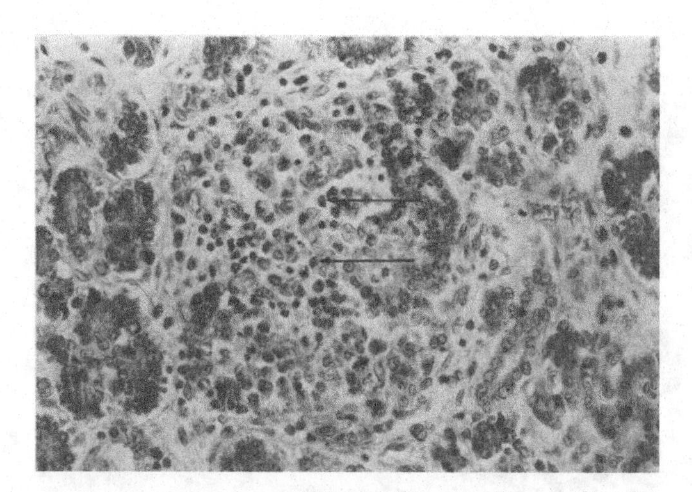

Fig. 10. Insulitis with many polymorphonuclear cells (arrows) is seen in neonate with fatal untyped Coxsackievirus B infection. (Reprinted by permission from ref. 2).

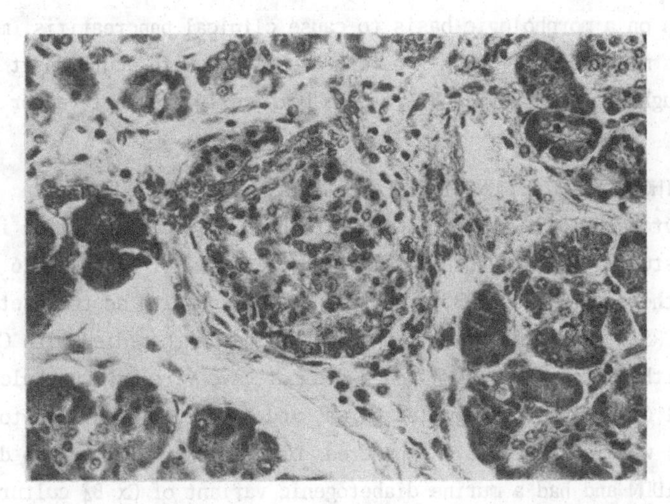

Fig. 11. Degeneration and necrosis with lysis of islet cells is characterized by indistinct nuclear and cytoplasmic boundaries in same neonate with Coxsackievirus B_4 infection as Figs. 8 and 9. (Reprinted by permission from ref. 2).

The distribution of viral lesions in the pancreas depends to a certain extent on the virus. CMV, V-Z, and Cx B affect both islet and acinar cells, usually with greater involvement of islets than acinar cells. In infants with culture-proven Cx B 1,2,4 and 5 viremia, islet cell damage was generally mild to moderate, varying from clusters of cells with pyknotic nuclei to necrosis of occasional islets, a pattern apparently characteristic of Cx B infections (4, 5). All islets are not affected equally. When present, inflammatory cells include polymorphonuclears, macrophages, and lymphocytes surrounding individual islets, although most islet cell necrosis has no associated inflammatory response. Cytomegalic cells of CMV appear in the islet cells, pancreatic ducts, and periacinar cells, almost always without associated necrosis. V-Z is characterized by inclusion bodies and prominent areas of necrosis involving both islet and acinar cells. In fatal acute infections with insular changes, all types of islet cells including beta and alpha cells are usually affected. Although mumps has been associated with pancreatitis and IDDM, there was only one case of fatal mumps infection among the autopsy reports studied by Jenson, et al. (1), but there were no pathologic changes in the pancreas. None of the pathologic changes in the pancreas of the 31 fatal cases with insulitis and pancreatitis appeared severe enough on a morphologic basis to cause clinical pancreatitis, much less diabetes mellitus (although two were associated with acute-onset IDDM) which is thought to occur after 80% of the beta cells are depleted or destroyed.

VIRAL CYTOPATHOLOGY AND IMMUNOCYTOCHEMICAL STAINING FOR INSULIN

Viral cytopathology was correlated with beta cell degranulation in those cases studied immunocytochemically for insulin. Of 12 cases, 6 had viral cytopathology (3 had Cx B, 1 had CMV associated with acute-onset IDDM, 1 had CRS 7 years after acute-onset IDDM and another died with CRS associated with acute-onset IDDM) and 6 did not have viral cytopathology (1 Cx A and 1 Cx B, 1 herpes simplex virus, and 2 CMV). In addition to the 150 cases (1) we also examined the pancreas of a 10-year-old boy who developed acute-onset IDDM and had a murine diabetogenic variant of Cx B_4 cultured from the pancreas (Fig. 2). IDDM was not documented in those children with clinical evidence of overwhelming disseminated viral infections. Conversely, the consequences of viral infection were neither a prominent clinical or pathologic feature in those children who died with concurrent viral infection and acute-onset IDDM.

The children who died with fatal viral infections and insular changes but without IDDM had varying degrees of degranulation of beta cells. This was particularly obvious in Cx B infections, where beta cell degranulation appeared segmental, involving partial or complete islets (Figs. 12-14). Many of the degranulated beta cells in Cx infections appeared to be shrunken and inactive, but still viable. The greatest degree of degranulation was approximately 60% in one of the patients with Cx infection. In patients with CMV and V-Z infection the degree of degranulation correlated fairly well with the extent of viral cytopathology in the islets (Fig. 15). In patients with fatal viremia but no pancreatitis, the islets apparently had their full complement of insulin-producing beta cells although some were partially degranulated. Thus the Cx infections had a much greater extent of beta cell degranulation than the other virus infections when islet cell insulin content was correlated with the degree of cytopathologic changes in the islets.

Of the four cases of viral infections associated with IDDM, two 10-year-olds with acute-onset IDDM and CMV and Cx B (6) involvement of the islets had approximately 20-30% and 50% of beta cells degranulated, respectively (Fig. 16). These findings suggest a functional loss rather than an actual loss of insulin in these two individuals, who would not be expected to have acute-onset IDDM on the basis of at least 50% of insulin capacity of islets theoretically available for glucose metabolism. On the other hand, the 14-month-old boy with rubella syndrome (7) and the 10-year-old with IDDM of 7 years duration and CRS had no insulin-producing beta cells in the islets. The absence of beta cells in CRS with IDDM correlates with the reports relating complete loss of insulin-producing beta cells to exolain the loss of insulin patients with IDDM. This may suggest a different mechanism for IDDM associated with rubella than IDDM associated with Cx B or CMV.

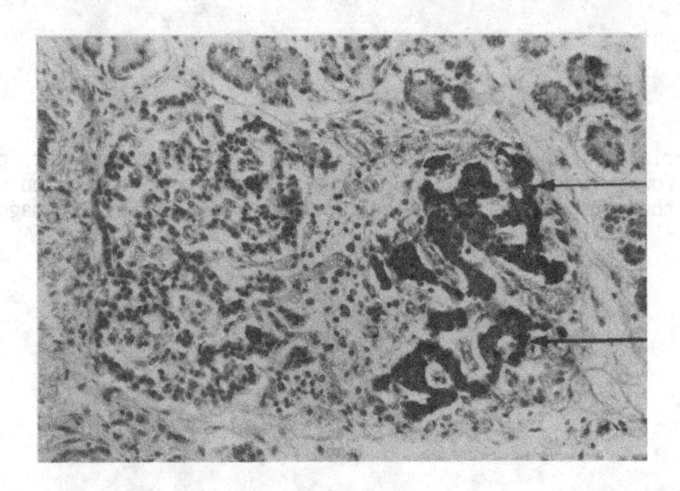

Fig. 12. Large islet bisected into segments by mixed inflammatory cell infiltrate in neonate with untyped Coxsackievirus B infection. This islet was stained for insulin by peroxidase anti-peroxidase (PAP) technique with segment of islet on right fully granulated (arrows) and segment on the left containing few granulated beta cells. (Reprinted from ref. 1).

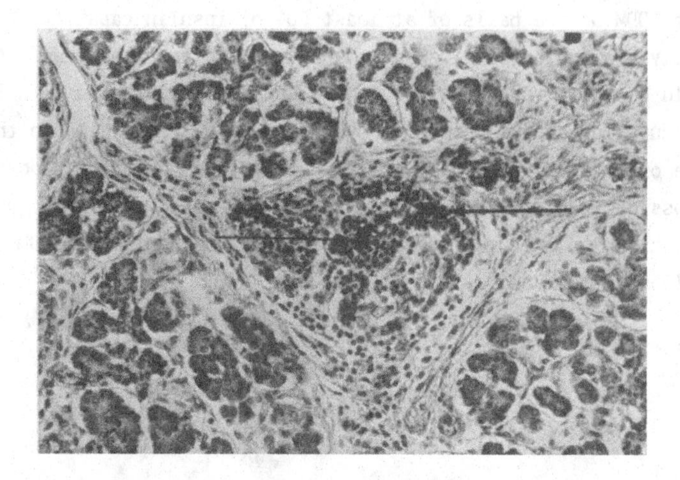

Fig. 13. Serial section of islet seen in Fig. 8 except stained for insulin granules (arrows) by PAP technique. Only a few beta cells are granulated even though there is minimal morphologic evidence of islet cell damage.

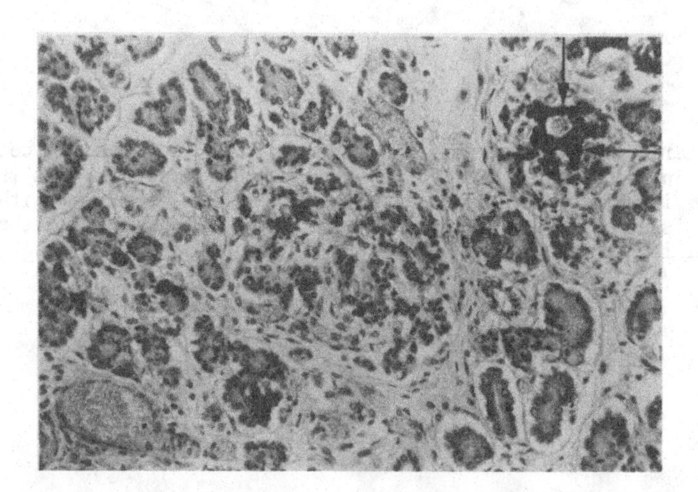

Fig. 14. Islet in center of field is mostly degranulated of insulin (PAP stain) whereas islet in upper right corner appears to be full granulated (arrows) in child with untyped Coxsackievirus B infection.

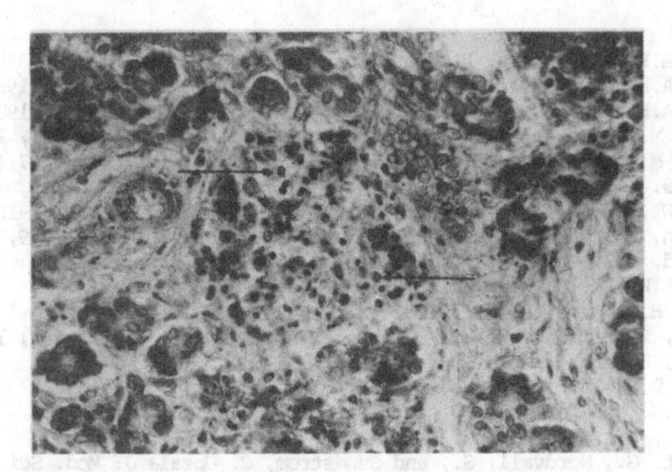

Fig. 15. Islet is partially degranulated of insulin (PAP stain) in neonate with congenital V-Z infection. Degranulation occurs primarily in beta cells with V-Z cytopathology (arrows) in contrast to Coxsackievirus B infections where degranulation often occurs in islet cells without evidence of degeneration and necrosis.

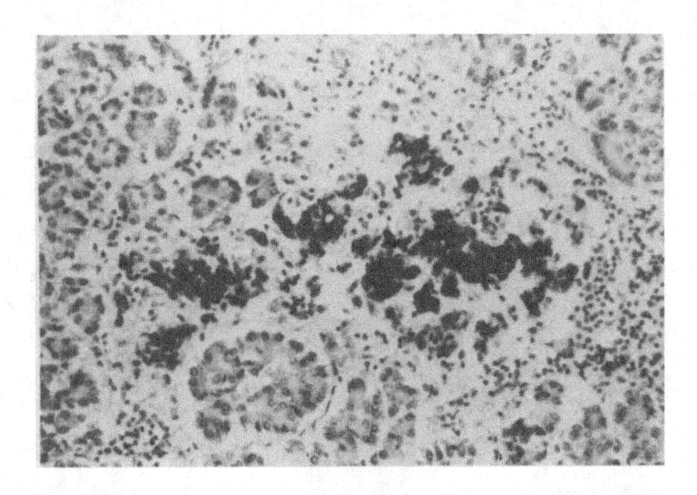

Fig. 16. Insulitis consisting of small lymphocytes surrounding and infiltrating islet containing granulated beta cells (PAP stain for insulin) in 10-year-old boy with Coxsackievirus B infection and diabetic keto-acidosis. Less than half of beta cells are degenerating and degranulated.

REFERENCES

1. Jenson, A.B., Rosenberg, H.S. and Notkins, A.L. Lancet ii:354-358, 1980.
2. Jenson, A.B. and Dobersen, M.J. Prog. Ped. Pathol. 7: 167-183, 1982.
3. Jenson, A.B. and Rosenberg, H.S. Prog. Med. Virol. 29:197-217, 1984.
4. Ujevich, M.M. and Jaffe, R. Arch. Pathol. Lab. Med. 104:438-441, 1980.
5. Sussman, M.L., Strauss, L. and Hodes, H.L. Am. J. Dis. Child. 97:483-492, 1959.
6. Yoon, J.W., Onodera, T. and Notkins, A.L. J. Exp. Med. 148:1068-1080, 1978.
7. Patterson, K., Chandra, R.S. and Jenson, A.B. Lancet i:1048-1049, 1981.
8. Craighead, J.E. Prog. Med. Virol. 19:161-214, 1975.
9. Farnam, L.W. Am. J. Med. Sci. 163:859-870, 1922.
10. Johnson, H.N. Arch. Path. 30:292-307, 1940.
11. Gladisch, R, Hofmann, W. and Waldhen, R. Z. Kardiol. 65:837-849, 1976.
12. Kalfayan, B. Arch. Pathol. 44:467-476, 1947.
13. Cappell, D.F. and McFarlane, M.N. J. Pathol. Bact. 59:385-398.
14. Smith, M.G. and Vellios, F. Arch. Pathol. 50:862-884, 1950.
15. Worth Jr., W.A. and Howard, H.L. Am. J. Pathol. 26:17-36, 1950.
16. Jultquist, G., Nordvall, S., and Sundstrom, C. Upsala J. Med. Sci 78:139-144, 1973.
17. Craighead, J.E., Kanich, R.E. and Kessler, J.B. Am. J. Pathol. 74:287-294, 1974.
18. Gepts, W. Diabetes 14:619-633, 1965.
19. Gepts, W. In Secondary Diabetes, Podolsky, S., and Viswanathan, M. (Eds.) (New York:Raven Press), pp. 15-32, 1980.

10

COXSACKIE B4-INDUCED PANCREOPATHY

SALLIE S. COOK, M.D., Assistant Professor of Pathology
ROGER M. LORIA, PH.D., Associate Professor of Microbiology and Pathology

Medical College of Virginia, Richmond, Virginia, USA

ABSTRACT

Supporting the hypothesis that Coxsackie B4 (CB4) may cause pancreopathy and/or be important in the pathogenesis of human diabetes mellitus are virologic, serologic, immunologic, genetic and pathologic studies. The most direct evidence for this relationship is isolation of the virus from a human pancreas showing histopathologic change consistent with the patient's diagnosis, juvenile (type 1) diabetes mellitus. Other laboratory assays, including measurement of islet cell antibodies and CB4 antibodies in populations of patients with juvenile diabetes mellitus provide further evidence for viral etiology of this disease. In the context of epidemiologic study, these and other observations, including age, sex and seasonal variation have been assessed. Additionally, positive correlations between certain human leukocyte antigen (HLA) determinants and viral antigenic responses, including CB4, have been found in studies of diabetic individuals.

INTRODUCTION

A temporal relationship between the onset of some viral infections and subsequent development of diabetes mellitus has been recognized since the turn of the century (1). It is generally accepted that there is hereditary predisposition to diabetes, however, there is little agreement on its mode of inheritance. The influence of viruses and other environmental factors on expression of these genotypes as clinical diabetes mellitus is uncertain (1-4).

Juvenile insulin - dependent (type 1) and adult onset non-insulin-dependent (type 2) diabetes are thought to be distinct entities, based on epidemiologic, genetic and immunologic studies (5). The concordance rate for diabetes in studies of identical twins has

Becker, Y (ed), Virus Infections and Diabetes Melitus. Publishing, Boston. ISBN 0-89838-970-4.

been reported to be 92% in adult-onset diabetes as compared to 53% in juvenile diabetes (6). While these observations indicate a strong genetic influence on the occurrence of adult-onset diabetes, the discordant results in juvenile diabetics suggest environmental factors may be important in the etiology of this disease (2).

The hypothesis that viruses are important in the pathogenesis of juvenile diabetes is supported by its seasonal incidence, abrupt clinical onset and accumulating serologic, immunologic and pathologic studies of patients with the disease (1,7,8,9). Implicated viruses include mumps, rubella, Coxsackie B, encephalomyocarditis virus, Epstein-Barr virus and cytomegalovirus (1,6,7,8,10). The most direct explanation for viral pathogenesis of the diabetic syndrome is infection and ultimate destruction of the insulin-producing beta cells of the islets of Langerhans in the pancreas, however, other hypotheses include altered immunoresponsiveness in the presence of viral antigens triggering autoimmune destruction of beta cells (11,12). Additionally, relationships have been noted between certain HLA serotypes and immunocompetence in the presence of CB4 and other viruses (13). Similar to viral oncogenesis, it has also been postulated that viruses may function as initiators of disease but may not be present at the time of disease expression (14). This later suggestion would obviously explain how a virus might be part of the etiology of a disease when all traces of the virus are absent at the time of diagnosis.

Animal studies have been performed to help elucidate the association between diabetes and viruses. CB4 has been implicated in studies including isolation of the virus from a ten-year-old patient with newly diagnosed diabetes mellitus. The diabetogenic potential of this virus was supported by mouse inoculation with the human isolate and subsequent demonstration of hyperglycemia and pancreopathy including beta-cell necrosis and inflammation in the islets of Langerhans (9).

CB4 and other picornaviruses have been further studied in various inbred strains of mice to assess their diabetogenic potential. As with human studies, multiple factors appear to influence susceptibility to viral-induced pancreopathy and abnormal glucose tolerance. Pancreatic damage related to these viruses and resulting in hyperglycemia appear to be dependent on genetic factors, (15,16,17), hormonal factors (15,18,19) and immunologic competency (20,21,).

This chapter addresses pertinent data from these animal models as well as human data supporting the relationship of CB4 and pancreatic pathology as it pertains to the clinical disease, diabetes mellitus.

CLINICAL DATA

Several clinical observations in patients with insulin-dependent diabetes mellitus (IDDM) suggest an acute infectious insult. In newly diagnosed cases, onset of symptoms is frequently abrupt. Also, many patients are previously healthy children who lack a positive family history of diabetes mellitus. In some cases there is a prior medical history of a viral infection.

One suggestion has been that diabetic members of a family have had less previous experience with Coxsackie viral infections in infancy and consequently had a more severe disease with Coxsackie B infections contracted when they were older. A study supporting this hypothesis showed that diabetic children had IgG titers >4 to a lower number of Coxsackie B 1-6 serotypes than age-matched nondiabetic siblings (22). It has also been shown that there is a more heterotypic IgM response to Coxsackie B 1-6 as nondiabetic children become older and have virologically confirmed Coxsackie B infections (23). It does appear, therefore, that less prior exposure to Coxsackie viruses, in general, may predispose some individuals to develop a more homotypic serologic response and, perhaps, more severe illness possibly causing pancreatic damage and resulting in the clinical syndrome, diabetes mellitus.

Acute "viral like" illnesses have been commonly reported prior to development of diabetic symptoms (24). Onsets of new cases are recorded more frequently in the winter and late summer - early autumn months, suggesting an association with common epidemic viruses of childhood (25). This seasonal incidence has correlated directly with peak occurrences of enterovirus infections. Gamble recognized a seasonal pattern, with one major peak at 11 to 12 years and secondary peaks at 5 and 8 to 9 years (26). He suggested that these peaks might be related to school entry at 4 to 5 years, transfer to elementary school at 7 to 8 years, to secondary school at 11 to 12 years and that virus infection might explain these observations (26). Another study, by Gamble and Taylor, documented that the variation in incidence of new cases of IDDM was paralleled by a similar variation in the numbers of isolations of CB4 virus (27).

While IDDM predominately affects patients under 40 years of age, sex predilection appears variable, depending on race and geographic location (28). Constitutional factors, however, such as poor nutritional status or metabolic and psychic stresses may be important in determining individual's susceptibility to pancreatic beta cell injury due to viruses (2). Several reports document the occurrence of clinical pancreatitis and specific lesions of the islets in disseminated Group B Coxsackievirus infections in man (29,30). Although acute enteroviral pancreatitis is uncommon in infants, children and adults, it has been associated with Coxsackie B viruses, especially 1, 2, and 4 (31). Pancreatitis seen as a part of disseminated Coxsackie infections is most commonly seen in neonates or persons with congenital or acquired immunodeficiency (29). Of the Coxsackieviruses, CB4 has been implicated most often as diabetogenic in clinical studies of IDDM (29).

These clinical observations have stimulated investigators to perform more indepth assessment of the virologic, immunologic and pathologic status of patients with IDDM.

PATHOLOGY

The first case of a patient having IDDM followed by isolation of a CB4 virus was reported by Yoon et al in 1979 (9). Following three days of flu-like symptoms, this ten-year-old boy was hospitalized in diabetic ketoacidosis and died seven days later (9). His flu-like illness consisted of fatigue, headache, sore throat, pleuritic chest pain, anorexia, nausea and myalgia. On admission, he was experiencing polyuria, polydipsia and abdominal cramps with increasing lethargia. The patient had chickenpox and mumps several years earlier but no recent drug or chemical exposure (9). On post-mortem examination, sections of the patient's pancreas showed lymphocytic infiltrates, often perivascular, surrounding the islets of Langerhans. This inflammatory infiltrate penetrated reticular fibers around the islets and extended into cellular columns of the islets. The cells of the islets were described as "often depeleted... and degenerating". Additionally, mild patchy chronic inflammation was noted throughout acinar tissue (9).

Lesions in the pancreas of persons with insulin-dependent diabetes mellitus are notoriously variable and no particular finding is believed to be pathognomonic (32,33,34). The following alterations are described: (1) reduction in the size and number of islets (2) increase in the size of islets (3) beta-cell degranulation (4) glycogen vacuolation of beta cells (5) hyalin replacement of islets or (5) leukocyte infiltration of islets (32,33). The most common pattern of leukocyte response is a heavy lymphocytic infiltrate within and around islets (32). Overall, the most common finding is gross reduction of islet tissue (32,24). In 1965, Gepts reported a description of the pancreas in patients dying within days or weeks of abrupt onset diabetes mellitus. All showed reduction in the number of islets and 15 of 22 (68%) had infiltration of islets by lymphocytes (35).

In our laboratory, a fifty percent lethal dose (LD_{50}) of CB4 (Edwards strain) has been used to infect three inbred strains of mice with subsequent assessment of the histopathology of their pancreas and other organs. This particular strain of CB4 was originally isolated from myocardial tissue of an infant with a generalized Coxsackie infection and focal necrosis and inflammation of the pancreas (36). When C57BL/6, SWR/J and DBA/2 mice were examined, histopathologic changes varied. By twenty-one days postinfection, 100% of SWR/J mice showed islet atrophy, acinar necrosis and a marked neutrophilic, lymphocytic and monocytic interstitial infiltrate with extension to peri-islet areas. In DBA/2 mice, sixty-six percent of animals showed a miminal interstitial inflammation (neutrophils, lymphocytes and monocytes) with occasional lymphocytic infiltration around the islets. C57BL/6 mice did not manifest any abnormal pancreatic pathology (37).

Interestingly, the diabetogenic effect of the CB4 strain used in our study also varied by measurements of glucose levels (37). C57BL/6 mice demonstrated abnormal glucose tolerance testing, analogous to chemical diabetes, despite the lack of pancreatic lesions on light microscopy. In contrast, SWR/J and DBA/2 strains showed increased glucose tolerance seven days post-infection. These latter results are similar to the so called "remission" or "honeymoon" phase of juvenile diabetes mellitus when glucose tolerance improves and, in fact, hypoglycemia may ensue as insulin requirements fall after the initial diagnosis of disease (38).

These anatomic and clinical laboratory findings in mice demonstrate that CB4 susceptibility as determined by LD_{50} values and pancreatic pathology can be independent parameters and do not necessarily determine the host glucose tolerance response of a given genotype. These studies, as with human data, suggest that specific host factors, i.e. genetics, hormonal factors, immunologic competency, may be of major importance in determining the diabetogenic potential of a specific viral agent (37).

Further studies by Loria et al have measured CB4 (Edwards strain) viral replication in numerous tissues of inbred mice with various genotypes as follows: C57BL/KsJ with diabetic (db) and misty coat (m) mutations, C57BL/6J with obese diabetic (ob), yellow obese (A^Y) and black (a) mutations and C57BL/Ks, an inbred control genotype. Using LD_{50} as infective doses, the CB4 concentrations in blood, pancreas, liver, spleen, heat and aorta were measured 3 days after viral challenge. Within each genotype, virus concentrations varied considerably in all tissues except the pancreas, where viral titers were consistently highest with a mean concentration of 9.54 ± 0.1 log PFU/gr tissue ($p < 0.001$). Based on this study, it was concluded that neither specific gene mutations (db, m, ob, A^y, a) nor background genotype, 6J and KsJ, markedly affected CB4 titers in various tissues; however, this study emphasizes the pancreatopic nature of this virus (39).

IMMUNOLOGY

Some human cases of IDDM appear to result from direct and massive infection involving the pancreas such as the patient reported by Yoon et al in 1979 (9). It now appears, however, that viral pathogenesis may be more frequently explained as one of several complex immune-related mechanisms. Both IgG and IgM antibody titers directed against various Coxsackie viruses as well as islet cell antibodies and their possible crossreactivity have now been measured in insulin-dependent diabetic patients.

In addition to documentation of the histopathologic lesions in the pancreas of the child reported by Yoon et al in 1979, there also was a demonstrable four-fold rise in the CB4 antibody titer (9). Prospective assessment of CB4 titers in another case report showed a greater than four-fold rise (1:129 to 1:512) three years prior to the clinical onset of insulin-dependent diabetes (40). Interestingly this latter patient also had a 1:16 titer of islet cell antibodies three years prior to the diagnosis of diabetes. Unfortunately, earlier specimens documenting the first occurrence of these antibodies were not available, thus, their temporal relationship is unknown.

A study by Orchard et al. of over one hundred new cases of IDDM using immunofluorescent staining methods for IgM (41) and IgG (42) showed that 34% had serological evidence of recent Coxsackie B 1-6 virus infection at diagnosis (IgG titer > 32 or IgM titer > 4). The control group for this study consisted of 57 siblings close in age to the diabetic group with results showing 42% had evidence of recent Coxsackie B infection (22). Significant, however, was the observation that the diabetic children were positive to a lower number of Coxsackie B 1-6 serotypes than the age-matched siblings.

Virus-specific IgM levels against Coxsackie B 1-6 have been assayed in several studies as a marker of current or recent antigenic stimulus in patients with recently diagnosed IDDM. Of these patients with elevated antibody titers, specificity is most commonly directed against CB4 (23,43,44). King et al in 1983 studied the presence of IgM against Coxsackie B viruses in 28 children with IDDM using an ELISA technique (23). They found IgM in 11 (39%) children, ten of whom had specificity against CB4. In contrast, Coxsackie B virus-specific IgM responses were seen in only 16 of 290 (5.5%) non-diabetic children serving as age-matched controls. Banatvala et al in 1985, also using an ELISA method, found Coxsackie B IgM responses directed against a single serotype in 23 of 37 (67%) children with IDDM, the most common serotypes being Coxsackie B4 (11/37) or B5 (11/37). Age-matched nondiabetic controls in this study showed 6% (13/204) positivity in Coxsackie B virus IgM levels (43). A third study reported by Frisk et al in 1985, detected Coxsackie B virus-specific IgM responses in 16 of 24 (67%) patients on the day of diagnosis of IDDM. This study, employing a reverse radioimmunoassay, showed monotypic responses in 13 of 16 patients, the largest percentage of these directed to CB4 (4/13) and the others to Coxsackie B 1,2,3 and 5. No Coxsackie B virus-specific IgM was detected in age-matched nondiabetic children in this latter report.

It is well accepted now that the majority of insulin-dependent diabetic patients have islet cell antibodies (ICA) (40,45,46). With respect to viral pathogenesis of IDDM, one proposed mechanism has been that the implicated virus and islet cells may share antigenic determinants; antibodies formed in response to the virus, therefore, might cross-react with the islet cells. Further support for this hypothesis comes from a preliminary study by Richens, Quilley and Hartog reporting that islet cell fluorescence of diabetic serum from three patients was decreased by prior absorption with Coxsackie B4 virus (11). Additionally, Schernthaner, Ludwig, and Mayr noted a significantly higher frequency of ICA in diabetics with moderate or high ($>$ 1:64) CB4 titers than those with low or undetectable levels of CB4 antibodies (47).

Arguing against the hypothesis that pancreatic islet cells and Coxsackie viruses share some antigenic determinants are several studies showing no correlation between CB4 and ICA titers when comparing pre-infection and post-infection samples from newly diagnosed diabetics (45,48,49). There are, however, other potential mechanisms whereby these viruses might be related to ICA. Cellular antigens might be incorporated into the viral surface during replication such that antibodies might be produced that would crossreact with islet cells and the viruses. Also, it is suggested that viral damage to islet cells might expose cytoplasmic antigens with resultant development of ICA (45).

In our laboratory, studies of inbred C57BL/KsJ mice, known to possess db and m mutations, have been performed to assess the host's immune response to CB4 challenge (6). Using the Cunningham-Szenberg plaque assay procedure (50), db/db infected or uninfected mice with overt disease demonstrated 100% elevation in spleen IgM counts ($p <$

0.01) while db/m infected mice had a 64% lower IgM cell count than uninfected mice ($p < 0.01$). Additionally, infected db/db mice failed to produce CB4 neutralizing antibodies while db/m and m/m mice developed high antibody levels 7 and 14 days after challenge. Thus, while this overtly diabetic animal model has intact spleen IgM-producing cells, significant CB4 antibody titers are not produced. This suggests that genetic predisposition to diabetes and the metabolic effects of overt disease alter the host's immune response to CB4 challenge (6).

It is clear, however, that mechanisms whereby viral infections are involved in an autoimmune component of IDDM remain to be explored.

GENETICS

It is generally accepted that there is a hereditary predisposition to diabetes (51). The heterogeneity of genetic origins giving rise to diabetes is demonstrated in studies analyzing the inheritance patterns of insulin-dependent and non-insulin-dependent forms of diabetes. Although a familial tendency to IDDM has been noted, less than 20 percent of these patients have first-degree relatives with a diabetic history and family studies have failed to show a definable pattern of transmission. Autosomal recessive inheritance with variable penetrance has, therefore, been most widely accepted (2).

Studies of monozygotic twins have been performed to assess the influence of genetics in diabetes. Categorizing patients by age, three reported studies have seen > 90 percent concordance among twins over 40 years of age. Contrary to this, approximately 50 percent concordance was observed in twins less than age 40 years (52,53,54). While this data supports a strong genetic influence on the occurrence of maturity-onset diabetes, environmental factors (including viruses) are felt to be important in juvenile-onset diabetes, possibly precipitating disease in genetically predisposed individuals (2).

A positive association of IDDM with HLA-DR3 and -DR4 has been observed in many Caucasoid populations, while a negative association with DR2 is noted (55,56). Bruserud, Jervell and Thorsby recently studied the relationship between responses to viral antigens (mumps, CB4 and varicella-zoster) and HLA-DR3 and -DR4 association in patients with IDDM (13). Antigen-reactive T-lymphocyte (ARTL) responses were measured in 8 insulin-dependent diabetics and 14 healthy individuals carrying DR3 or DR4 HLA determinants. With both CB4 and Mumps, DR3-associated hypo-responsiveness and DR4-associated hyper-responsiveness were seen in both diabetic and healthy individuals. No DR-associated change in response was found with varicella zoster in any subjects (13). These results are felt to be consistent with the hypothesis that certain viruses may be important in the etiology of IDDM. The lower frequency of ARTL to CB4 and mumps in association with DR3 may render individuals carrying DR3 more vulnerable to infection of beta cells with these viruses. Further human studies, however, will be necessary to clarify the immunoregulatory functions of these and other HLA determinants with respect to the viral immune response.

Extensive work by Loria et al has been performed on inbred C57BL/KsJ mice to determine whether genetic predisposition to diabetes mellitus and/or clinical diabetes influence the production of neutralization antibodies (NA) to CB4 (57,58). Antibody titers were measured pre-infection, 3-21 days and 1-5 months after LD_{50} infection of CB4 in C57BL/KsJ mice with m and db mutations. Control mice (m/m, db/m) showed consistent NA elevations beginning at 3 days post-infection and reaching maximum titers of > 90% plaque reduction at 7 days post-infection. In contrast, mice homozygous for the diabetic mutation, db, had no detectable NA titer until after 2 months. Homozygous db/db mice with no restricted food intake developed significant hyperglycemia (glucose = 693 mg/dl 2 hours after challenge of 2 mg glucose/g body weight) during glucose tolerance testing and NA levels of 90% virus plaque reduction by 3 months post-infection. Homozygous db/db mice with diet restriction did not develop overt diabetes and CB4 NA titers were low (approximately 20% plaque reduction), short-lived and probably represented a non-specific response. Further study of the C57BL/KsJ db/db mice that did not develop overt diabetes with food restriction has shown that their deficient humoral immunity involves general impairment in both total IgM and IgG production after CB4 infection (59). These mice in which phenotypic expression of diabetes is prevented by diet restriction suggest that immune impairment may be found prior to diabetes onset when compared to db/db mice with overt diabetes.

Both this animal data and the previously cited human data (13) suggest significant difference in immune response to CB4 among genotypes with variable diabetic phenotypic expression.

CONCLUSION

This chapter summarizes recent human data addressing the relationship of CB4 and pancreopathy which may be demonstrated clinically as diabetes mellitus and anatomically with histopathologic lesions in the pancreas. Clinical evidence, including serologic, immunologic and genetic data, have been presented which suggest etiologic involvement of CB4 in IDDM. Epidemiologic studies have shown seasonal incidence, occurrence of acute "viral-like" illness prior to diabetic symptoms and prevalence of viral antibodies in IDDM patients as compared to normal controls. Experimental animal models with varying genetic backgrounds have been infected with CB4 and add evidence that this viral-induced pancreopathy may, indeed, result in diabetes mellitus.

It is clear, however, that while compelling evidence exists to support CB4-induced pancreopathy resulting in IDDM, the exact etiologic role of this virus or other external influences, as they relate to genetic factors, is not yet confirmed. Further development of experimental models and elucidation of pathophysiologic events that occur with human coxsackie viral infections will be required for better interpretation of these relationships.

REFERENCES

1. Notkins, A.L. Arch. Virol. 54:1, 1977.
2. Craighead, J.E. N. Engl. J. Med. 299:1439, 1978.
3. Yoon, J.W., Onodera, T., Notkins, A.L. J. Exp. Med. 148:1068, 1978.
4. Gamble, D.R., Taylor, K.W., Cumming, H. Br. Med. J. 4:260, 1973.
5. National Diabetes Data Group. Diabetes, 28:1039, 1979.
6. Loria, R.M., In: Lessons from Animal Diabetes (Eds. Shafrir, E., Renold, A.E.). John Libbey and Co., 1984.
7. Craighead, J.E., Prog. Med. Virol. 19:161, 1975.
8. Rayfield, E.J., Seto, Y. Diabetes 27:1126, 1978.
9. Yoon, J.W., Austin, M., Onodera, T., Notkins, A.L. N. Engl. J. Med. 300:1173, 1979.
10. Craighead, J.E., Human Path. 10:267, 1979.
11. Richens, E.R., Quilley, J., Hartog, M. Acta. Diabetal. Lat. 15:229, 1978.
12. Marx, J.L. Science 223:1381, 1984.
13. Bruserud, O., Jervell, J., Thorsby, E. Diabetologia 28:420, 1985.
14. Huppert, J., Wild, T.F. Ann. Virol. 135:327, 1984.
15. Boucher, D.W., Hayaski K., Rosenthal J., Notkins, A.L. J. Infect. Dis. 131:462, 1975.
16. Craighead, J.E., Higgins, D.A. J. Exp. Med. 139:414, 1974.
17. Yoon, J.W., Notkins, A.L. J. Exp. Med. 143:1170, 1976.
18. Craighead, J.E., Steinke J. Am. J. Pathol. 63:119, 1971.
19. Maugh, T.H. Science, 188:436, 1975.
20. Buschard, K., Rygaard, J., Lund, E. Pathol. Microbiol. 84:299, 1976.
21. Jansen, F.K., Munterfering, H., Schmidt, W.A.K. Diabetologia 13:545, 1977.
22. Orchard, T.J., Atchison, R.W., Becker, D., Rabin, B., Eberhardt, M., Kuller, L.H., LaPorte, R.E., Cavender D. Lancet ii:631, 1983.
23. King, M.L., Bidwell, D., Shaikh, A., Voller, A., Banatvala, J.E. Lancet ii: 1397, 1983.
24. John, H.J. J. Ped. 35:723, 1949.
25. Adams, S.F. Arch. Intern. Med. 37:861, 1926.
26. Gamble, D.R. Proc. R. Soc. Med. 68:256, 1975.
27. Gamble, D.R. and Taylor, K.W. Br. Med. J. 3:631, 1969.
28. West, K.W. In: Epidemiology of Diabetes and Its Vascular Lesions, Elsevier, New York, 1978, pp. 216-221.
29. Sussman, M.L., Strauss L., Hodes, H.L. Am. J. Dis. Child. 97:483, 1959.
30. Gladisch R. Hofmann, W., Waldherr R.Z. Kardiol. 65:837, 1976.
31. Moore, M., Morens, D.M. In: Textbook of Human Virology (Ed. R.B. Belshe) PSG Publishing Company, Inc., Littleton, 1984, pp. 456-458.
32. Robbins, S.L., Cotran, R.M. In: Pathologic Basis of Disease, W.B. Saunders Co., Philadelphia, 1979, pp. 336-338, 1095-1096.
33. Warren, S., Le Compte, P.M. Legg, M.A. In: The Pathology of Diabetes Mellitus, Lea and Febiger, Philadelphia, 1966, pp. 53-115.
34. Doniach, I. Proc. R. Soc. Med. 68:256, 1976.
35. Gepts, W. Diabetes, 14:619, 1965.
36. Kibrick, S., Benirschke K. Pediatrics 22:857, 1958.

37. Cook, S.H. S., Loria, R.M., Madge, G.E. Lab. Invest. 46:377, 1982.
38. Sperling, M.A. In: Nelson's Textbook of Pediatrics (Eds. Behrman, R.E., Vaughan, V.C., Nelson, W.E.), W.B. Saunders Co., Philadelphia, 83, pp. 1414-1415.
39. Loria, R.M., Montgomery, L.B., Corey, L.A., Chinchilli, V.M. Arch. Virol. 81:251, 1985.
40. Asplin, C.M., Cooney, M.K., Crossley, J.R., Dornan, T.L., Roghu, P., Palmer, J.P. J. Ped. 101:398, 1982.
41. Henle, W., Henle G.E. Hum. Pathol. 5:551, 1979.
42. French, M.L.V., Schmidt, N.J., Emmon, R.W., Lennette, E.H. Appl. Microbiol. 23:54, 1972.
43. Banatvala, J.E., Schernthaner, G., Schober, E., DeSilva, L.M., Bryant, J., Barkenstein, M., Brown, D., Menser, M.A., Silink, M. Lancet ii: 1409, 1985.
44. Frisk, G., Fohlman, J., Kobbah, M., Ewald, U., Tuvemo, T., Diderholm, H., Friman, G. J. Med. Virol. 17:219, 1985.
45. Palmer, J.P., Cooney, M.K., Crossley, J.R., Hollander, P.H., Asplin, C.M. Diabetes Care 4:525, 1981.
46. Srikonto, S. Garda, O.P., Eisenbarth, G.S., Soeldner, J.S. N. Engl. J. Med. 308:322, 1983.
47. Schernthoner, G., Ludwig, H., Mayr, W. R. Acta. Diabetol. Lat. 15:184, 1978.
48. Irvine, W.J., Al-Khateeb, S.F., DiMario, U., Feek, C.M., Gray, R.S., Edmond, B., Duncan, L.J. Clin. Exp. Immunol. 30:16, 1977.
49. Cudworth, A.G., White, G.B.B., Woodrow, J.C., Gamble, D.R., Lendrum, R., Bloom, A. Lancet i:385, 1977.
50. Cunningham, A.J. and Szenberg, A. Immunol. 14:599, 1968.
51. Renolds, A.E., Stauffacker, W., and Cahill, G.F. In: The Metabolic Basis of Inherited Disease (Eds. W.W. Foster, J.L. Goldstein, J.L. Brown, M.S. Stonburg, J.B. Wyngaarden, D.S. GredErickson), McGraw-Hill, N.Y., 1983, pp. 99-117.
52. Pyke, D.A., Nelson, P.G. In: The Genetics of Diabetes mellitus (Eds. W. Cruetzfelch, J. Kobberling, J.V. Neel). Springer-Verlog, Berlin, 1976, pp. 194-202.
53. Then Berg, H. Gesamte Nearol. Psychiatr. 165:278, 1939.
54. Gottleib, M.S., Root, H.F. Diabetes 17:693, 1968.
55. Svejgaard, A., Platz, P. Ryder, L.P. Immunol. Rev. 70:193, 1983.
56. Scholz, S., Albert, E. Immunol. Rev. 70:77, 1983.
57. Loria, R.M., Montgomery, L.B., Tuttle-Fuller, N., Gregg, H.M., Chinchilli, V.M. Diab. Res. Clin. Prac. 2:91, 1986.
58. Loria, R.M., Behring. Inst. Mitt. 75:26, 1984.
59. Montgomery, L.B., Loria, R.M. J. Med. Virol 19:255, 1986.

53. [illegible], [illegible] Microbiol. [illegible] 153:275, 1986.

11

DIABETES MELLITUS ASSOCIATED WITH EPIDEMIC INFECTIOUS HEPATITIS IN NIGERIA

F.C.ADI

Bethsaida Clinic, Box 2211 Enugu, Nigeria.

ABSTRACT

This report concerns nine cases of diabetes mellitus associated with infectious hepatitis, an epidemic of which swept through eastern Nigeria between 1970 and 1972 following the Nigerian civil war. All the patients suffered from classical acute infectious hepatitis and, as they appeared to be recovering from this infection, developed symptoms and signs of diabetes mellitus. They responded quickly to treatment and after a few months, the diabetes clinically disappeared. Corticosteroid-glucose tolerance tests carried out in four of these patients, 12 to 30 months after the clinical remission of their diabetes, were normal. Contact with the remaining five patients was lost after a few months follow-up following clinical remission, probably because they remained well. It is postulated that the virus that caused the acute infectious hepatitis may have damaged pancreatic islet cells, partially or temporarily, to produce an acute remittant form of diabetes mellitus in these patients.

INTRODUCTION

An association between virus infections and diabetes mellitus is well documented in both animals and man. Diabetes mellitus has followed an attack of foot-and-mouth disease in cattle (1,2). The encephalo-myocarditis (EMC) virus, especially the M variant, has been shown to cause pancreatic islet tissue damage, with the development of a diabetes-like syndrome in mice (3,4,5).

Becker, Y (ed), Virus Infections and Diabetes Melitus. ISBN 0-89838-970-4.

Coxsackie B4 and B1 virus infections have also been shown to cause pathological changes in pancreatic islet tissue of suckling mice by Burch and his colleagues (6) and this could be associated with hyperglycaemia, usually after a lag phase of 2-3 weeks (7).

The possibility that a virus might cause diabetes mellitus in man was first suggested in 1899 by Harris (8). Since that time, a number of virus infections have been reported to be associated with the onset of diabetes mellitus. The virus most often incriminated has been the mumps virus. Cole (9) suggested this association in three young diabetics. Meling and Ursing (10) reported four cases following an epidemic of mumps in Sweden when they studied forty children and two adults. Hinden (11) found reports of 20 cases of diabetes mellitus following mumps and added a case of his own where severe diabetes mellitus with coma occurred in a child, about five weeks after an attack of mumps. Another virus often suspected is the rubella virus. Forrest et al (12) reported that in 44 patients with congenital rubella, approximately 20 percent ultimately developed diabetes mellitus or showed an abnormal glucose tolerance test. Other viral infections reported to have been associated with diabetes mellitus have included those caused by cytomegalovirus, tick-borne encephalitis virus, Coxsackie B4 and Venezuelan equine encephalitis viruses. Our present experience in Nigeria seems to indicate that the virus that causes classical epidemic infectious hepatitis (Hepatitis A virus) or a close relation of it, can also produce biochemical as well as clinical diabetes mellitus.

PATIENTS AND METHODS

The nine patients who developed clinical diabetes mellitus during an attack of acute infectious hepatitis form part of a group of about 100 such patients seen by the author during an epidemic of viral hepatitis,which ravaged the eastern part of Nigeria from 1969 to 1972.The epidemic started during the dying stages of the Nigerian civil war, in what was then known as "the Republic of Biafra". An estimated seven to eight million people were crammed into a land space of about 60 by 40 miles (80 by 64 Km) and besieged for over two years, in what must have been one of the longest sieges in recent history. With a break-down in public health measures and severe overcrowding, a classical epidemic of infectious hepatitis swept through a population already severely ravaged by starvation and malnutrition. This background picture might be pertinent in that the nutritional status of the patients may have played some role in the development of the diabetes complication. Nevertheless, it is important to note that the civil war ended in January 1970, but the epidemic of hepatitis continued until 1972, and all the nine patients reported in this paper were seen between 1970 and 1972, when the nutritional status of the people as a whole had considerably improved.

All the nine patients presented with classical symptoms and signs of diabetes mellitus, with marked wasting and dehydration associated with polydipsia and polyuria and in some cases with keto-acidosis. They had glycosuria and a raised fasting blood sugar. They all required insulin followed by oral hypoglycaemic drugs for some months. They then clinically remitted and were followed up for at least 12 months.

During follow up, the patients were made to test their urine daily with Clinitest tablets for the first four to eight weeks and thereafter on alternate days. At first no sugar or glucose was allowed but otherwise the normal African diet, consisting mostly of carbohydrate foods such as yam, cassava, maize and rice was unrestricted. The reason for not restricting the diet was that practically all the patients when first seen were wasted both from the acute hepatitis, with its associated anorexia, and from the acute diabetes. On a free diet and insulin, the diabetes was quickly brought under control, and since the need for insulin and oral hypoglycaemic drugs progressively diminished, there was no need to interfere with the patients' usual diet, following withdrawal of all drugs.

Patients were recalled after being followed up, and remained completely healthy, for varying periods ranging from 12-30 months. Five patients defaulted after being followed up for less than 12 months and so could not be re-evaluated. The remaining four patients were each subjected to a modified corticosteroid-glucose tolerance test using the method of Fajans and Conn (13). Each patient was given a total of 40 mg of prednisolone in four divided doses on the day before he reported as an outpatient for a standard glucose-tolerance test. Blood glucose level was estimated on capillary blood samples by the method of Folin and Wu.

CASE REPORTS

Case 1

The patient, a 39-year old printer, was first seen and admitted to hospital in December 1970 with polyuria, polydipsia and loss of weight over the preceding two weeks. There was no family history of diabetes mellitus and no past history of any diabetic episode or glycosuria. He had taken ill in September 1970 with severe infectious hepatitis. An epidemic of jaundice had been raging in the area where he lived at the time. Someone had advised him to take a lot of glucose for the hepatitis and he had taken a total of 39 Kg of glucose powder over the preceding four months. Sometime in September, he had been given a short course of prednisolone-15 mg of prednisolone daily for a week. On examination, he was slightly jaundiced and moderately emaciated. The urine contained over 2% glucose but no ketone bodies. Fasting blood sugar was 250 mg/dl.

He was started on soluble insulin 20 units three times daily just before each meal. Ordinary African type ward diet was allowed but without any sugar or glucose. The diabetes was quickly controlled and the insulin requirement progressively reduced. After two weeks, he was discharged on chlorpropamide 500 mg daily, reducing to 250 mg daily after a few weeks follow-up. All anti-diabetic therapy was stopped in February 1971. For the next 14 months, despite an unrestricted diet, he remained healthy and his urine tests were consistently free of glucose. He was then recalled and a corticosteroid-glucose tolerance test carried out, which was quite normal. The patient when last seen in January 1986, sixteen years later, was quite healthy and fully at work.

Case 2

The patient, a 36-year-old electrician, was first seen in January 1971 complaining of severe weight loss, marked weakness, severe polyuria and polydipsia for about three weeks. There was no family history of diabetes mellitus. He had been off work for the past eight weeks because of severe jaundice, an epidemic of which was raging at the time. He had received a total of 15 mg of prednisolone daily for three weeks, taken a total of about 13 Kg of glucose powder for the hepatitis, and was recovering steadily when the diabetic symptoms appeared.

On examination, he was very emaciated, extremely weak and grossly dehydrated. His urine contained over 2% glucose and ketone bodies. There were no facilities for emergency blood glucose estimation and the patient refused admission to hospital. He was treated with 3 litres of isotonic saline intravenously and 40 units of soluble insulin twice daily as an outpatient. He recovered slowly but steadily over the next two weeks and his insulin requirements fell progressively to 40 units daily at the end of this period. He was then switched to chlorpropamide 500 mg daily, reducing to 250 mg daily over the next fourteen weeks. He was allowed to eat normally except for sugar. He gained weight from 51 Kg to his usual weight of 62 Kg and remained well and his urine glucose-free. All anti-diabetic drugs were stopped on 29 May 1971. Nineteen months later, a corticosteroid-glucose tolerance test was normal. When seen more than two years after stopping all anti-diabetic treatment, the patient was well and at work.

Case 3

The patient, a 47-year-old businessman when first seen on 23 August 1971 had been ill for two weeks with malaise, vague joint pains, marked tiredness, anorexia and dark-coloured urine. He had been treated for malaria but was not improving. He was jaundiced but otherwise looked well. His urine contained much bilirubin and excess urobilinogen but no glucose. He weighed 86 Kg. A diagnosis of infectious hepatitis was made, reinforced by a history of recent contact with the disease. He improved rapidly with treatment which included prednisolone 15 mg daily for three weeks. He regained his strength and appetite and his jaundice had almost completely disappeared when he began to notice polyuria, polydipsia and rapidly lost weight despite his good appetite. After two weeks of suffering from these symptoms, he again sought medical advice.

There was no family history of diabetes mellitus. He looked emaciated, his weight having dropped to 70 Kg. The urine contained more than 2% glucose and a trace of ketones. Liver function tests were almost normal with a total bilirubin of 1.4 mg/dl and conjugated fraction of 0.5 mg/dl. A standard glucose tolerance test the day after his admission showed a fasting blood sugar of 280 mg/dl and a classical diabetic curve with the blood sugar remaining as high as 240 mg/dl two and half hours after a 50 g glucose load.

He was treated as usual with soluble insulin followed later by chlorpropamide and was discharged after three weeks stay in hospital. He returned to his normal business eating normally but avoiding sugar and alcohol. Four months later all anti-diabetic drugs were stopped as patient had remained well and his urine sugar free. His weight had risen to 82 Kg. After an uneventful follow-up for 18 months, he was recalled and a corticosteroid-glucose tolerance test was done and found to be entirely normal. When last seen in 1985, fourteen years after the above event, he was quite well and had no evidence of diabetes mellitus.

Case 4

The patient, a 28-year-old civil servant, was first seen on 12 May 1972. Seven weeks previously he had developed classical acute infectious hepatitis with a positive contact history. He had been given prednisolone 15 mg daily by his doctor for two weeks. After about six weeks when his jaundice had cleared and he was almost fully recovered, he developed classical symptoms of diabetes mellitus with glycosuria and a fasting blood sugar of 157 mg/dl. He was admitted to hospital and treated along the same lines as the other three patients reported above. He responded promptly to treatment and was discharged after only ten days in hospital on chlorpropamide 500 mg daily. This was progressively reduced and all anti-diabetic drugs discontinued after about seven weeks treatment. The patient was followed up for one year during which time he remained healthy and eating whatever he liked except sugar. He was then recalled and a corticosteroid glucose tolerance test was carried out which was quite normal.

DISCUSSION

The aetiology of diabetes mellitus though not fully understood is now known to be multifactorial, embracing hereditary factors, insufficient production of insulin by pancreatic islets, overproduction of some other hormones, circulating insulin antagonists, production of abnormal forms of insulin and obesity. The role of excessive ingestion of carbohydrate foods, which must promote a sustained demand for large amounts of insulin, has not yet been fully eluciated. The present series of cases suggests that the role of infections of pancreatic islet tissue by certain viruses should be more vigorously considered.

Since my report of these nine cases many more such cases have been seen by both the author and some other Nigerian doctors (14,15). Oli and Nwokolo (15) for example reported seeing 11 patients during the same epidemic of acute infectious hepatitis who developed diabetes mellitus, one of whom remained permanently diabetic while the remainder remitted either completely or partially over a four year follow-up.

Questions we have to consider are: Was the causative virus of this epidemic of infectious hepatitis, the hepatitis A virus usually associated with such epidemics, or were we dealing with a more virulent strain of the same virus? Were we seeing in Nigeria an epidemic of classical acute infectious hepatitis, but being caused by a new type of hepatitis virus? Was this yellow fever epidemic misdiagnosed? Could this have been a hepatitis B virus, or hepatitis non A-non B virus-induced "epidemic"?

From the epidemiological point of view, we can assert that we had an epidemic of hepatitis that involved a very large number of people, running into thousands. We had a situation where overcrowding, poor sanitation and a polluted water supply set the stage for a "hand-to-mouth" borne epidemic infection, such as has been known to be caused by the hepatitis A virus. Hepatitis B and hepatitis non A-non B virus have not previously been reported to cause such widespread epidemics in the general population. Nevertheless Francis et al (16) did carry out some serological tests during the epidemic and showed that it was not due to hepatitis B virus. The clinical picture of the individual patients did not suggest Weil's disease nor a diagnosis of yellow fever. Furthermore, a World Health Organisation expert on yellow fever, who was called in by the State Ministry of Health, found no evidence of yellow fever virus as the cause of the hepatitis epidemic. From the clinical point of view, the illness ran the usual course as classically described for acute infectious hepatitis in all the standard text-books of medicine, <u>except</u> that in some of the patients, after several weeks of illness, and usually as they were recovering from the hepatitis, they developed symptoms and signs of diabetes mellitus.

Turning our attention to the diabetes, what could have been the cause? Were these patients potential or latent diabetics who were predestined anyway to develop diabetes mellitus sometime, and who were made clinically overt by the stress of infection? Were these cases of steroid-induced diabetes? Could it be that the insulin stores in the pancreatic islet tissue of these patients were temporarily exhausted by the prolonged demand of a high carbohydrate or glucose intake? Or could it be that the virus which caused the hepatitis also infected the pancreatic islet cells, preventing them from producing enough insulin and that healing resulted in restoration of their insulin-producing capacity?

That these were merely episodes of clinical diabetes mellitus in latent diabetics is unlikely because none of these patients had a family history of diabetes nor a previous history of glycosuria. Moreover when they were eventually subjected to a corticosteroid-glucose tolerance test, they were found to be normal. Neither can the mere stress of the virus infection be held primarily responsible, for the diabetes manifested itself several weeks after the onset of the hepatitis and usually as the patients were recovering. Thus the peak of the stress of the infectious hepatitis had passed before the onset of the diabetes. The diabetes is unlikely to have been due to the prednisolone given to some of the patients. The dose never exceeded 15 mg daily; it was given for not more than three weeks, and, apart from case 3, the diabetic symptoms appeared after the prednisolone was stopped. Moreover when these patients were later challenged with a corticosteroid-loaded glucose tolerance test, they showed no abnormality. We are therefore left with two possibilities: either the capacity of the pancreatic islet cells to produce enough insulin was temporarily exhausted by the stimulus of a sustained high carbohydrate intake or the capacity was diminished by actual infection of the islet cells by the hepatitis virus. We do not know what role the stimulus of a prolonged carbohydrate intake plays in the normal person as regards the possibility of exhaustion of the insulin stores. One would however expect that with the normally high carbohydrate diet of the average Nigerian, his pancreatic islet cells should have physiologically adapted to this kind of stimulus. In disease states however the same argument may not apply. Seltzer and Harris (17) maintained a prolonged stimulation of the pancreatic islets by infusing glucose intravenously for seven days in normal persons; in mild diabetics and in elderly insulin-dependent diabetics. The normal subjects showed greatly increased plasma insulin levels, retained all the infused glucose, and showed no hyperglycaemia. The mild diabetics showed an intermediate response, but the elderly patients showed a more pronounced and more persistent hyperglycaemia and glycosuria. More significantly, the plasma insulin level of these elderly diabetics rose only for the first few days of the infusion and then fell almost to zero. Thus it is possible under certain circumstances to stimulate pancreatic insulin secretion to a point of exhaustion.

Turning lastly to the possibility that the pancreatic islet cells may have been infected by the same virus that caused the hepatitis, one could not ignore the recurring 'coincidences' of acute diabetes mellitus occurring in a close temporal association with hepatitis in our patients, especially as there is abundant evidence in both animals and man that certain viruses can infect the pancreatic islet cells and produce diabetes mellitus (18). Although not usually reported as one of such viruses, the hepatitis A virus, or a close relation, should now be included in the growing list of such viruses (19). The evidence though purely clinical and epidemiological is overwhelming and the only question is why similar observations had not been previously noted in past epidemics of acute infectious hepatitis.

The latent period between the time of presumed infection of the pancreatic islets and the appearance of the diabetes is quite intriguing and may throw some light on the exact mechanism following infection, which causes the insulin deficiency. Viraemia occurs very early during the hepatitis infection and this is probably when the viruses infect the pancreatic islet cells. Presumably they replicate and cause damage to the islet cells just as they do in the liver cells.

In the vast majority of patients, the body defense mechanisms develop antibodies which successfully eliminate the viruses from the liver, and the patient makes full recovery usually within 6-8 weeks. One wonders what is happening to the viruses in the pancreatic islet cells at this time. Are they continuing to grow and damage the islet cells merrily immune to the antibodies against them, until sufficient damage has occurred to produce diabetes? Does an antigen-antibody reaction occur at this time within the pancreatic islet cells producing the damage which precipitates the acute diabetes? Or is the acute diabetes a manifestation of auto-immune disorder where the virus damages the islet cells and the cellular material acts as an antigenic stimulus leading to the elaboration of islet-cell antibodies which produce further damage to remaining islet cells? Interesting and intriguing as these questions are, they fall outside the scope of this article. However it is pertinent to note that this latent period had been noted in animal experiments by Martin and Lacy (20) who found that after surgical removal of 95% of pancreatic tissue of rats, a prediabetic phase of about three weeks occurred before the diabetes. In cases of diabetes mellitus in man following mumps infection this latent period has lasted from a few weeks to nine months (10,11).

The patients reported here showed latent periods of one to three months and remained clinically diabetic from between six weeks and five months. Clearly whatever type of damage their pancreatic islets suffered it was only partial or temporary. The possibility that some less fortunate patients may have suffered a more permanent damage cannot be ruled out.

REFERENCES

1. Pedini, B., Avellini, G., Moretenni, B., Comodo, N.: Att. Soc. Ital. Sci. Vet. 16, 443-450, 1962.
2. Barboni, E., Manocchio, I.: Arch. Vet. Ital. 13, 477-489, 1962.
3. Boucher, D., Notkins, A.L.: J. exp. Med. 137, 1226-1239, 1973.
4. Craighead, J.E., Steinke, J.: Amer. J. Pathol. 63, 119-130, 1971
5. Hayashi, K., Boucher, D.W., Notkins, A.L.: Amer. J. Pathol. 75, 91-102, 1974.
6. Burch, G.E., Isui, C.Y., Harb, J.M., Colcolough, H.L.: Arch. Int. Med. 128, 40-47, 1971.
7. Coleman, T.J., Taylor, K.W., Gamble, D.R.: Diabetologia 10, 755-759, 1974.
8. Harris, H.F.: Boston Med. and Surg. J. CXL 465-469, 1899.
9. Cole, L.: Lancet 1 947, 1934.
10. Meling, K., Ursing, B.: Nordisk Medicin 60, 1715, 1958.
11. Hinden, E.: Lancet i, 1381, 1962.
12. Forrest, J.M., Menser, M.A., Burgess, J.A.: Lancet ii 332-334, 1971.
13. Fajans, S.S., Conn J.W.: Diabetes 3, 296, 1954.
14. Adi, F.C.: Br. Med. J. 1, 183-185, 1974.
15. Oli, J.M., Nwokolo, C.: Br. Med. J. 1, 926-927, 1979.
16. Francis, T.I., Smith, J.A., Wright, S.G.: Tropical and Geographical Medicine, 24, 44, 1972.
17. Seltzer, H.S., Harris, V.L.: Diabetes, 13, 6, 1964.
18. Notkins, A.L.: Arch. Virol. 54, 1-17, 1977.
19. Ganda, P., Soeldner, S.S.: Arch. Intern. Med. 137, 461-469, 1977.
20. Martin, M., Lacey, P.E.: Diabetes 12, 238, 1963.

12

HUMAN VENEZUELAN EQUINE ENCEPHALITIS AND DIABETES

Elena Ryder and Slavia Ryder

Instituto de Investigaciones Clínicas, Facultad de Medicina, Universidad del Zulia, Maracaibo, Venezuela.

This chapter reviews the effect of Venezuelan equine encephalitis (VEE) virus on pancreatic beta cells. Experimental studies have been done in several animals and it has been shown that the virus is capable of attacking beta cells. In hamsters, the Trinidad strain produced a massive infection and death occurred very shortly after inoculation; mature virions and viral antigens were found in the beta cells (1). Three months after inoculation of the VEE-attenuated strain TC-83 there was no visible damage to the islets but metabolic studies showed some abnormalities in glucose tolerance tests, although the fasting glycemia levels were not elevated. The insulin content of each pancreas tested was normal but the insulin response to glucose was occasionally depressed (1). The same pattern was described in monkeys by Bowen et al. (2). These investigators reported that in young (2-3 year old) rhesus monkeys infected with the virulent Trinidad donkey strain of VEE virus, the glucose tolerance curves 2, 5 and 10 months after infection were abnormal and that insulin curves showed a progressive diminution of both, basal and glucose-stimulated insulin levels. A normal pancreatic histology and insulin content in these animals were compatible with a defect in insulin release from beta cells. However in a more controlled study in older monkeys (3-8 years old) done by the same authors (2) no pancreatic dysfunction due to epizootic VEE virus or live attenuated VEE (TC-83) vaccine was observed. The discrepant results were assumed to be due to differences in the ages of the animals used in the two experiments.

In C57BL/Ks mice, db/db strain, the TC-83 strain produced complete degranulation of beta cells and an important metabolic alteration characterized by low levels of immunoreactive circulating insulin, low content of pancreatic insulin and an abnormal glucose tolerance test (3).

Although several epidemiologic studies have suggested a seasonal

Becker, Y (ed), Virus Infections and Diabetes Melitus. ISBN 0-89838-970-4.

incidence in the onset of the insulin-dependent diabetes in humans for certain viruses such as coxsackie, mumps, rubella and others (4, 5) there are no reports on the consequent metabolic status after a VEE epidemic.

Human epidemic outbreaks of VEE, confirmed by isolation of virus and/or presence of antibodies in sera, have occurred in several regions of the Western hemisphere: Venezuela, Trinidad, Colombia, Ecuador, Central America, Mexico and Southern United States (6).

In Venezuela the VEE virus produced frequent epidemic outbreaks in a cyclic manner in the Guajira peninsula of Zulia State, located in the Western part of the country (Fig. 1). The first report of some clinical cases was in 1959. In 1962 a large epidemic struck the Guajira with more than 30,000 cases and 190 fatalities. Three more outbreaks occurred in the same area in 1968, 1969 and 1973 (6).

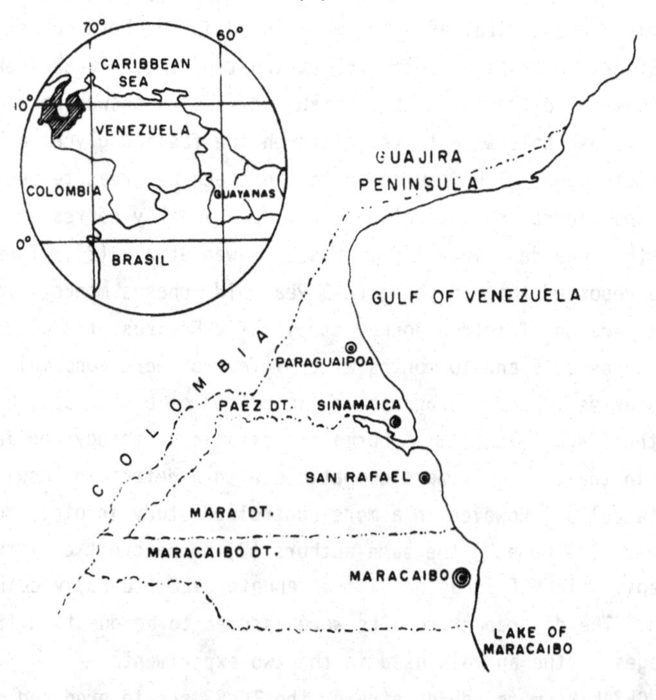

Fig. 1 The map shows the districts from Zulia State (located in the western part of Venezuela) more often affected by the VEE outbreaks between 1959 and 1973. The human studies reported here were performed in this geographical area.

The virus attacked mainly infants producing important neurological symptoms (6). Results from autopsies performed on 15 children (2-11 years old) from the 1962 epidemic, revealed intense hyperemia in all viscera (7) but there is no mention of specific pancreatic lesions. In none of the clinical or pathological descriptions of the different human cases is there any mention of metabolic alterations related to carbohydrate metabolism (6, 8). Follow-up studies on groups of patients who suffered from the infection during these epidemics in Venezuela (9) and Colombia (10) did not mention metabolic abnormality as a sequela of the disease.

Moreover, epidemiological data from the same geographical area showed no abnormal incidence of diabetes among the inhabitants of this particular zone of Venezuela (11) and the prevalence of carbohydrate intolerance seems to be the same as in the rest of the country where the presence of the virus has not been reported (12).

To get a better insight into the problem, in 1978 we performed a detailed metabolic study on thirteen patients who suffered VEE infection during the 1969 epidemic [when 1380 cases were confirmed by neutralization tests (13)] nine years after recovering from the infection, but still with persisting high titers of antibodies (40 to 640 HIA) (14). This seems to be the only metabolic study in VEE-infected humans reported in the literature.

Serum glucose and insulin immunoreactive levels (Fig. 2) in response to an oral glucose load revealed that although the shape of the curves were not strictly similar, we could not find any significant differences among them that indicated carbohydrate intolerance or an altered insulin response to glucose as described in hamsters (1) or monkeys (2). Almost 50% of the patients studied were less than 20 years old indicating that at the time of infection they were infants; the rest were over 45 years old. Hence, in humans we cannot implicate age as a predisposing factor for the development of pancreatic alterations as suggested by Bowen et al. for experimentally infected monkeys (2).

Another approach to find a link between VEE infection and the development of diabetes was based on a survey performed among 86 diabetic patients, ages ranging from 11 to 80 years. The presence of VEE viral antibodies in this group of patients was compared with that in a similar number of normal people, from the same geographical zone. Of the 86 diabetic patients, 9 (10.4%) had antibodies to the VEE virus detectable at 1:10 or greater dilution of serum, compared to 7 out of 98 controls (7.1%). This difference was not statistically significant. In addition, most individuals with antibodies developed diabetes after 40 years of age and were catalogued as non-insulin dependent or 'maturity onset' diabetics (14).

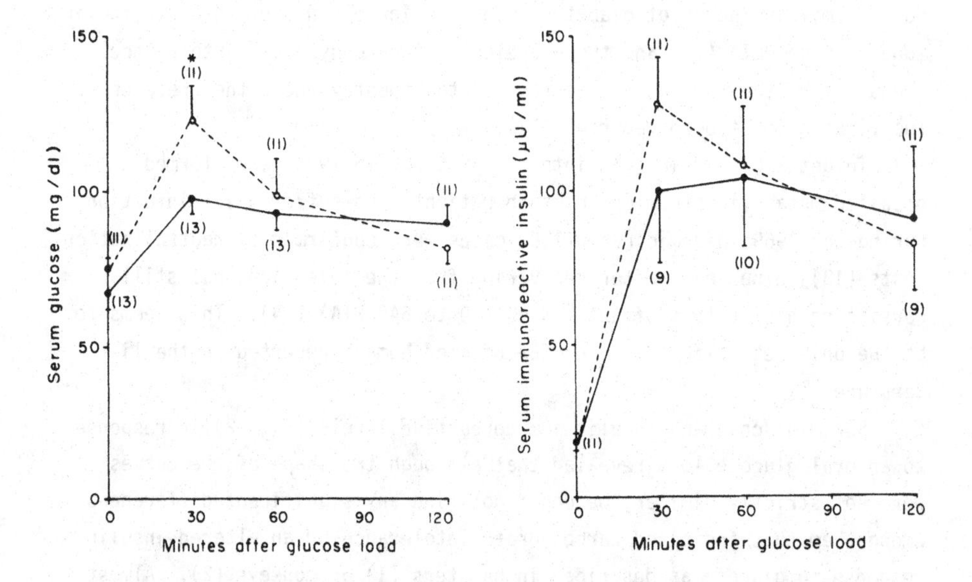

Fig. 2 Glucose and insulin responses to an oral glucose tolerance test in VEE-infected patients ●—● and controls Δ---Δ
The points in the curve represent the mean ± s.e.m. of the number of individuals shown in parenthesis. The asterisk indicates a $p < 0.05$ at the given time.

A selected study done on a group of children recognized and treated as type I diabetics (14) revealed the absence of antibodies to VEE virus indicating that in those particular cases, there was no participation of the virus in the development of the disease.

We have no data on the immunological pattern of the population in the Guajira peninsula and neighboring areas, but we cannot discard the possibility of the role of a genetic factor that controls the development of virus-induced diabetes in man.

The picture that emerges from the results reviewed is that VEE virus does not seem to produce a diabetic syndrome as a result of the infection in humans at least under the circumstances present after the outbreaks that occurred in South America.

REFERENCES

1. Rayfield, E.J., Gorelkin, L., Curnow, R.T. and Jahrling, P.B. Diabetes 25: 623-631, 1976.
2. Bowen, G.S., Rayfield, E.J., Monath, T.P. and Kemp, G.E. J Med. Virology 6: 227-234, 1980.
3. Rayfield, E.J., Sato, Y., Goldberg, S.L., Schulman, R.H. and Walker, G.F. Diabetes 28: 799-803, 1979.
4. Notkins, A.L. Arch. Virology 54: 1-17, 1977.
5. Maugh, T.H. Science 188: 347-351, 1975.
6. Ryder, S. Invest. Clin. 13: 91-141, 1972.
7. Wenger, F. Invest. Clin. 4: 23-45, 1963.
8. Garcia-Tamayo, J. Invest. Clin. 21: 277 - 371, 1980.
9. Castillo, C.E. Rev. Vlana. S.A.S. 31: 889-897, 1966.
10. Leon, C., Jaramillo, R., Martinez, S., Fernández, F., Tellez, H., Sasso, B. and de Guzman, R. Int. J. Epidem. 4: 131-140, 1975.
11. Ministerio de Sanidad y Asistencia Social. Memoria y Cuenta de la Dirección Regional de Salud, Región Zuliana. 1982-1983.
12. Semprun-Fereira, M., Ryder, E. and Campos, G. Invest. Clin 26 (Supl. 1): 50, 1985.
13. Ryder, S., Finol, L.T. and Soto-Escalona, A. Invest. Clin. 12: 52-63, 1971.
14. Ryder, E. and Ryder, S. J. Med. Virol. 11: 327-333, 1983.

Viruses as diabetes-causing agents in mice

13

VIRUS-INDUCED DIABETES IN MICE

JI-WON YOON

Division of Virology, Department of Microbiology and Infectious Diseases, and Laboratory of Viral and Immunopathogenesis of Diabetes, Julia McFarlane Diabetes Research Unit, The University of Calgary, Health Sciences Centre, Calgary, Alberta, Canada T2N 4N1.

ABSTRACT

Encephalomyocarditis (EMC) virus, Mengovirus 2T and Coxsackie virus B4 can induce diabetes in mice by infecting and destroying pancreatic beta cells. The destruction of beta cells in EMC virus-infected mice is not dependent on the immune responses but the genetic background of the host and genetic makeup of the virus. In addition to the acute metabolic alterations, mice with diabetes for six months show some long-term complications, including glomerulosclerosis, ocular changes and decreased bone formation and mineralization. EMC virus-induced diabetes in mice can be prevented by interferon and/or a live attenuated vaccine. In contrast to EMC induced diabetes, reovirus type I may cause a polyendocrine disease characterized by a mild and transient type of diabetes that can be prevented by immunosuppression. It is concluded that there are at least two different pathogenic mechanisms for virus-induced diabetes in mice: the one is direct cytolytic infection of beta cells and the other is the autoimmune mediated destruction of beta cells.

INTRODUCTION

Insulin-dependent diabetes mellitus (IDDM) results from the destruction of pancreatic beta cells. During the last several decades, genetic factors, autoimmunity and viral infections have been extensively studied as the possible cause of beta cell destruction. The evidence for virus-induced diabetes comes largely from experiments in animals (1-4), but several studies in humans also point to viruses as a trigger of this disease in some cases (2-8).

The best experimental evidence indicating that viruses have an etiological role in the pathogenesis of IDDM appears to be that of mice

Becker, Y (ed), Virus Infections and Diabetes Melitus. ISBN 0-89838-970-4.

infected with encephalomyocarditis (EMC) virus (1, 9). In genetically susceptible mice, the M-variant of EMC virus can infect and destroy pancreatic beta cells (1, 9, 10, 14) (Fig. 1). This results in a diabetes-like syndrome characterized by hypoinsulinemia, hyperglycemia, glycosuria, polydipsia, and polyphagia. The severity of the diabetes correlates closely with the degree of virus-induced beta cell damage (11). In addition to EMC virus, reovirus type I might cause mild and transient hyperglycemia through the triggering of autoimmune response (12). In this review, I would like to summarize some of our studies as well as others on virus-induced diabetes in mice.

I. Direct Cytolytic Infection of Beta Cells.

A group of viruses including EMC virus, Mengovirus and Coxsackie B viruses can directly infect murine pancreatic beta cells and replicate in the cells. The replication of viruses in the beta cells results in the destruction of the cells, and the infected mice subsequently develops hypoinsulinemia and hyperglycmia.

A. Encephalomyocarditis Virus: Encephalomyocarditis (EMC) virus, one of the smallest RNA viruses belongs to the picornavirus family (Fig. 2A). When this virus infects the susceptible cells, RNA and protein synthesis of the host are inhibited within 3 to 5 hours after infection. Work with EMC virus involves the most extensively studied animal model for virus induced diabetes.

1. Genetic Control of EMC Virus-Induced Diabetes: When mice were infected with the M-variant of EMC virus, only certain inbred strains of mice, such as SJL/J, SWR/J, DBA/1J and DBA/2J developed diabetes while other strains, such as C57BL/6J, CBA/J, AKR/J did not develop diabetes (1, 3, 13, 17). When diabetes-prone SWR/J mice were crossed with diabetes-resistant C57BL/6J mice, the F1 offspring were resistant to diabetes when infected with the virus (1, 3, 13). More than 20% of the F2 offspring, however, developed diabetes when exposed to the virus, indicating that susceptibility was inherited as an autosomal recessive trait. When the resistant F1 progeny were backcrossed to the resistant C57BL/6J parents, the offspring also were resistant to the development of EMC virus-induced diabetes. In contrast, when the resistant F1 progeny were backcrossed to the susceptible SWR/J parents, approximately one-half

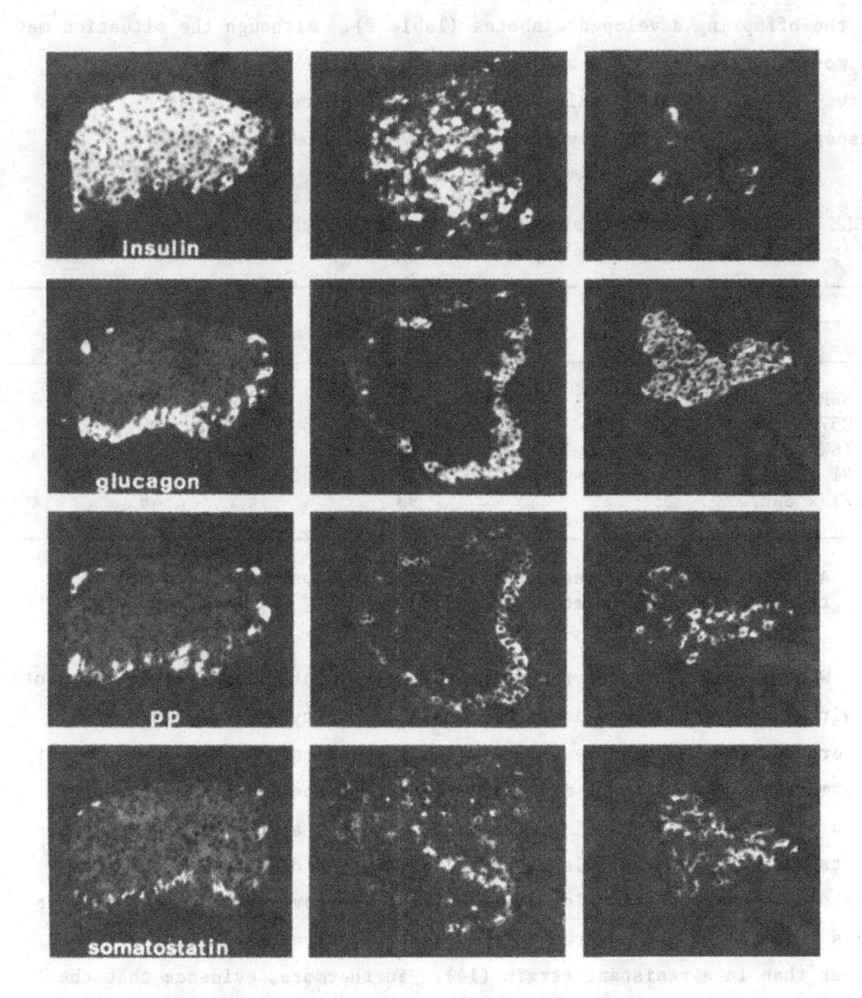

Fig. 1. Immunofluorescent reactions in serial section of pancreas of: (1) a control mouse (left panel): note the peripheral localization of the non-B-cells. (2) a hypoglycemic mouse 3 days after infection (middle panel): glucagon and PP-immunofluorescent cells are still located at islets periphery; several somatostatin immunofluorescent cells can be seen in the center of the islet, (3) a hyperglycemic mouse 21 days after infection (right panel): note the reduction of islet size, scarcity of insulin-immunofluoresent-cells and abundance of glucagon-immunofluorescent cells; PP-and somatostatin-immunofluorescent-cells are scattered throughout the islet (X 165).

of the offspring developed diabetes (Table 1). Although the situation may be more complex, the data are consistent with the idea that EMC virus-induced diabetes follows Mendelian inheritance and that susceptibility is primarily controlled by a single locus (13).

TABLE 1: Genetic Control of EMC Virus-Induced Diabetes*

Strains of Mice	Number of Mice	Diabetes (%)
SWR/J	33	88
C57BL/6J	33	0
(SWR/J x C57BL/6J)F1	45	0
F1 x C57BL/6J	70	5
F1 x SWR/J	93	46

* An animal was scored as diabetic if its glucose index was 3SD above the mean of 346 uninfected animals.

What is the nature of the genetic factors controlling the development of virus-induced diabetes in mice? One possibility is that host differences in the capacity of the virus to replicate in beta cells might determine which strains of mice would develop diabetes. Measurement of virus titer revealed that EMC virus replicated to a greater extent in the islets from mice susceptible to EMC virus-induced diabetes than in the islets from resistant strains of mice (14). Moreover, the number of beta cells infected with EMC virus in a susceptible strain was about 10 times higher than in a resistant strain (14). Furthermore, evidence that the severity of virus-induced diabetes is secondary to the degree of viral replication in islet cells from studies on the F1 and F2 offspring of susceptible and resistant strains of mice (15). These studies showed that blood glucose and plasma insulin levels in EMC virus-infected F1 mice remained within the normal range, and that virus titers in islets of F1 mice were similar to those found in islets of resistant mice. In contrast, F2 mice segregated into two groups, one group with plasma insulin and blood glucose levels within the normal range and the other group with low plasma insulin levels and high glucose levels. Analysis of individual F2 mice revealed two groups: one with viral titers approaching the susceptible mice and the other with titers similar to those of the resistant mice.

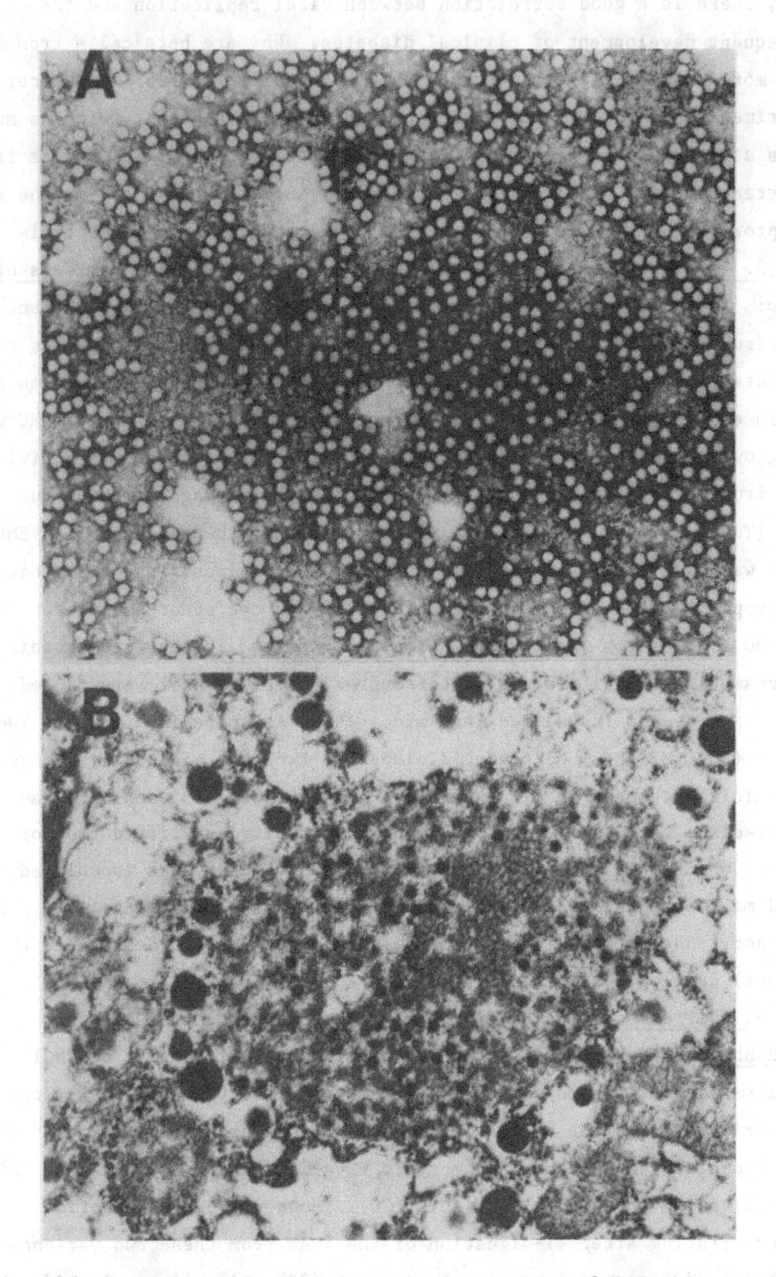

Fig. 2. (A) Electron microgram of purified D variant of EMC virus particles prepared by CsCl gradient centrifugation. The size of the virion is approximately 25 nm in diameter; (B) electron micrograph of reovirus particles in pancreatic beta cells of mice 5 days after infection. The size of the virus is approximately 75 nm in diameter.

Thus, there is a good correlation between viral replication and the subsequent development of clinical diabetes. Why are beta cells from SWR/J mice more susceptible to infection that beta cells from C57BL/6J mice? Experiments measuring viral attachment showed that at least twice as much virus attached to the pancreatic beta cells from susceptible mice as from resistant strains of mice (16). These data suggest that there may be more receptors for EMC virus on the surface of beta cells from susceptible mice.

2. Isolation of the Diabetogenic and Non-diabetogenic Variants of EMC Virus: The development of diabetes after infection with the M-variant of EMC virus was not consistent. Statistically significant differences were consistently found upon repetition of the same experiment and between cages within experimental groups (17). Moreover, when the M-variant of EMC virus was grown in mouse embryo fibroblast cultures, the diabetogenic activity of the virus markedly diminished in contrast to the passage of the virus in mice (Table 2) (18). These findings suggest that the stock pool of EMC virus was made up of at least two populations of virus: one diabetogenic and tropic for insulin-producing beta cells and the other non-diabetogenic with no tropism for insulin-producing beta cells. To see whether this was the case, individual plaques selected from the stock pool were cloned several times and inoculated into mice (19). As shown in Figure 3, Two clones (Nos. 82 and 108) produced diabetes, two (Nos. 5 and 16) produced only minimal changes in blood glucose, and two (Nos. 125 and 162) gave intermediate results. Clones 108 and 16 were plaque-purified two more times. When clone 108-D-II (designated the D variant) was inoculated into SJL/J male mice, diabetes developed in more than 90% of the animals. In contrast, none of the mice inoculated with clone 16-B-I (designated the B variant) developed diabetes (Fig. 4).

3. Differences and Similarities Between the Diabetogenic and Non-diabetogenic Variants of EMC Virus: The buoyant density on CsCl density gradient and capsid polypeptides on polyacrylamide gels of these two variants could not be distinguished. Molecular hybridization studies with radiolabeled DNA complementary to D variant and B variant RNAs failed to distinguish the D and B variants (20). However, oligonucleotide finger-printing after T1-digestion of the RNAs from these two variants revealed a difference in at least one spot (20). An oligonucleotide about 25 bases long was missing from the B variant but present in the D variant (Fig. 5).

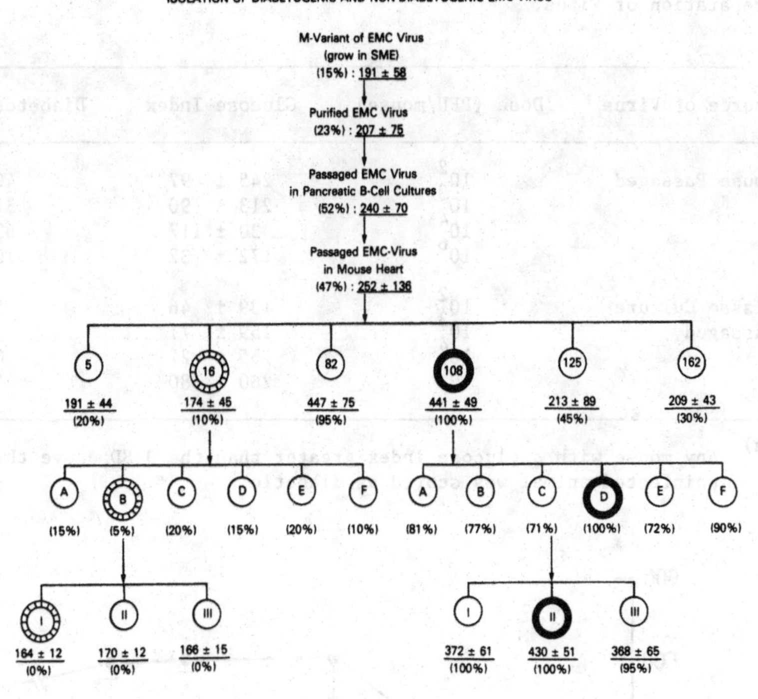

Fig. 3. Isolation of diabetogenic and nondiabetogenic variants of EMC virus. The M variant of EMC virus was passaged 5 times in murine pancreatic beta-cell cultures. The virus was then passaged 5 additional times in SJL/J male mice. The virus was serially plaque-purified. SJL/J male mice were inoculated with virus (5 x 10^5 PFU) and blood glucose was measured. The mean glucose index of 110 uninfected SJL/J male mice was 145 ± 19 mg/dl. Any mouse with a glucose index greater than 240 mg/dl, which was 5 SD above the mean, was scored diabetic. "( )" represents percent diabetes and "—— " represents mean ± SD of glucose index. The data were compiled from 20 mice per group.

TABLE 2: Induction of Diabetes in Swiss NIH Mice with Different Doses and Preparation of Virus.

Source of Virus	Dose (PFU/mouse)	Glucose Index	Diabetes[a] (%)
Mouse Passaged	10^2	245 ± 97	40
	10^3	213 ± 90	31
	10^4	330 ± 117	63
	10^6	172 ± 32	10
Tissue Culture Passaged	10^2	139 ± 46	5
	10^3	159 ± 71	3
	10^4	165 ± 27	0
	10^6	260 ± 80	45

[a] Any mouse with a glucose index greater than the 3 SD above the mean of uninfected animal was scored as diabetic.

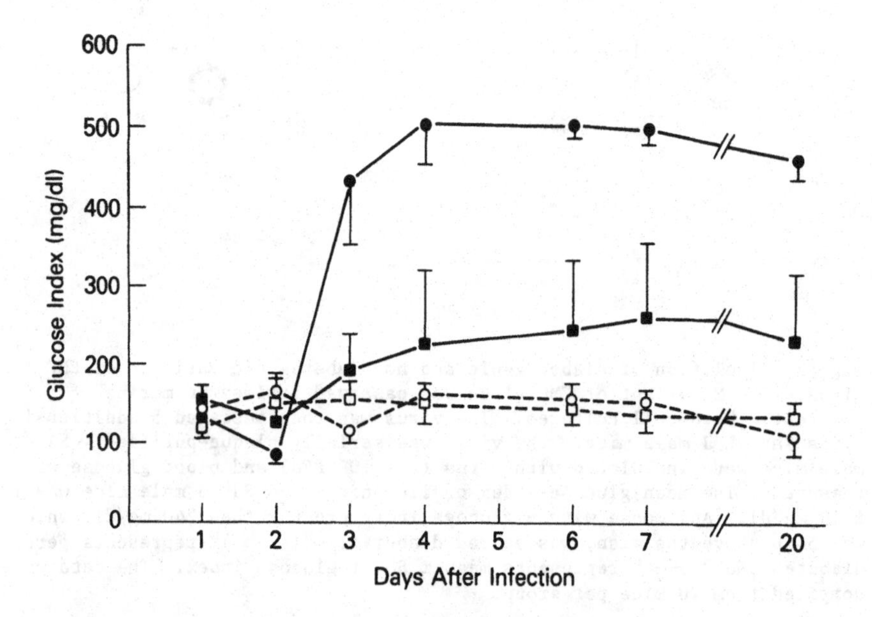

Fig. 4. Blood glucose levels of mice infected with the D or B variant of EMC virus. Each mouse received 10^6 PFU of virus. At the times indicated, the mice were bled, and nonfasting glucose (NFG) levels were determined. Each point represents the mean of ten animals; vertical bars are the SD. D variant (-●-), B variant (-○-), M variant (-■-), and uninfected control (-□-).

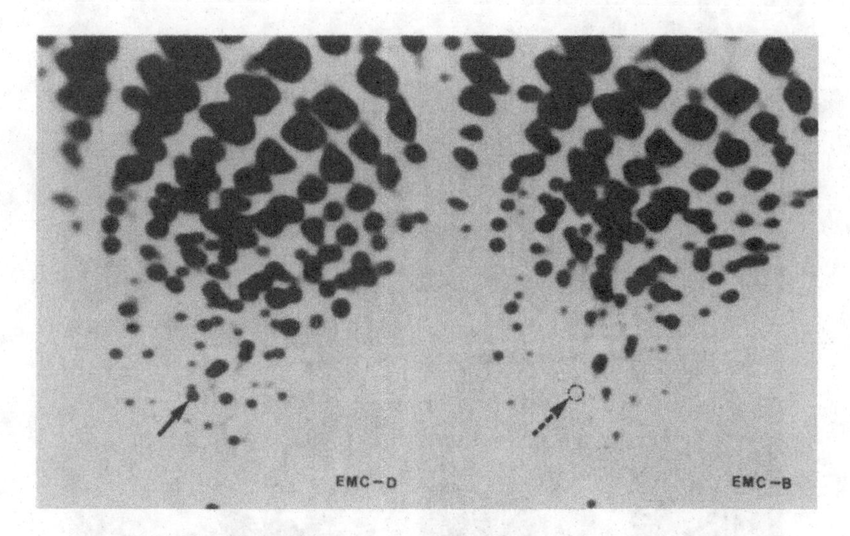

Fig. 5. Fingerprints of EMC RNA. Unlabelled RNA (2 to 3 µg) was digested with ribonuclease T1 and calf intestine alkaline phosphatase for 45 min at 37°C. The digested RNA was precipitated with ethanol after extraction of the reaction mixture with phenol-chloroform. The fragmented RNA was heated at 70°C for 3 min and then terminally labelled with [γ-^{32}P]ATP using T4 polynucleotide kinase. The labelled oligonucleotides were separated by two-dimensional gel electrophoresis. Solid arrow indicates one D-variant specific spot which contains about 25 nucleotides and dot arrow indicates the location which is missing the spot from B-variant.

Biological differences between these two variants are even more severe. Light microscopy revealed that the D variant, but not the B variant, produced severe beta cell damage (Fig. 6B) (19). Fluorescent microscopy using fluorescein-labeled anti-EMC antibody, showed that in mice inoculated with the D variant of EMC virus, approximately ten times more beta cells became infected than in mice inoculated with the B variant of EMC virus (Fig. 6A). Moreover, by measuring infectious virus, 10 to 100 times more virus was recovered from islets of mice infected with the D variant than from those with the B variant (19). Tissue culture experiments showed that the D variant induced little if any interferon, whereas substantial amounts of interferon were produced by the B variant (19).

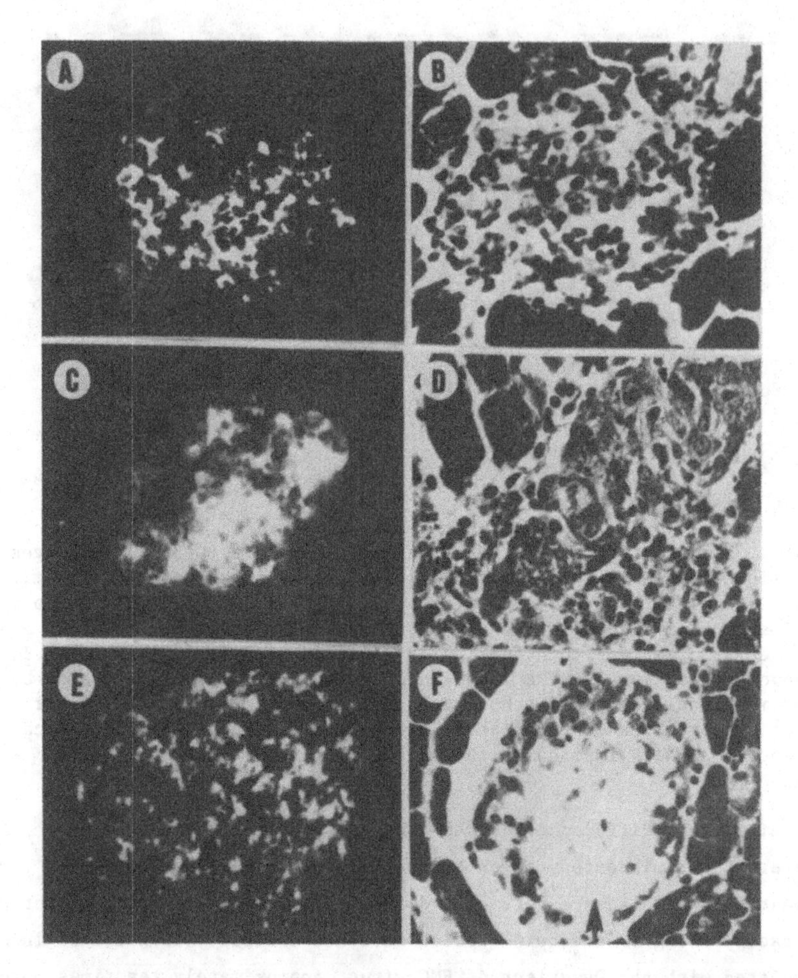

Fig. 6. Section of pancreas from mice infected with the EMC-D virus (A, B), Coxsackie B4 virus (C, D) or Mengovirus 2T (E, F). (A) Section obtained 3 days after infection with the EMC-D virus and stained with FITC-labeled anti-EMC virus antibody. The majority of the cells in the islets contain viral antigens. The surrounding acinar cells show little fluorescence (x 350). (B) Section taken 7 days after infection with EMC-D virus showing extensive inflammatory infiltrate in the islets of Langerhans and beta-cell necrosis. Hematoxylin-eosin stained (x 550). (C) Section obtained 4 days after infection with Coxsackie B4 virus and stained with FITC-labeled anti-Coxsackie B virus antibody. Most of the cells in the islet of Langerhans contained viral antigens. The surrounding acinar cells

are free from viral antigen (x 350). (D) Section of pancreas 5 days after infection with Coxsackie B4 virus showing extensive infiltrate of islet with lymphocytes (H and E x 500). (E) Section obtained 2 days after infection with Mengovirus 2T and stained with FITC-labeled anti-Mengovirus antibody. The majority of the cells in the islets contain viral specific antigens (x 550). (F) Section taken 6 days after infection with Mengovirus 2T showing complete loss of normal islet architecture and severe coagulation necrosis (x 550).

Despite these differences, the D and B variants could not be distinguished antigenically by a sensitive plaque neutralization assay (19). Antibody made against the D variant neutralized both the D and B variants. Conversely, antibody made against the B variant neutralized the D and B variants equally well. Furthermore, competition radioimmunoassays also failed to reveal any major differences between the two variants (21). These studies illustrate the potential difficulty in identifying diabetogenic viruses in nature. Based on standard serologic tests, it would be difficult to distinguish the diabetogenic D variant from the non-diabetogenic B variant.

4. Mechanism for the Destruction of Pancreatic Beta Cells by the D Variant of EMC Virus: What is the mechanism involved in the destruction of beta cells? Is the destruction of beta cells due to the autoimmune response? Several investigators have suggested that immune mechanisms may be involved in EMC virus (M-variant)-induced diabetes. Jansen and coworkers reported that 500 R X-irradiation protected mice from virus-induced IDDM (22). Buschard *et. al.*, in experiments with two strains of nude mice, reported that these mice did not develop virus-induced diabetes, whereas their heterozygous littermates became diabetic (23, 24). Mortality was high in some of the experiments, and both groups of investigators infected mice with the M-variant of EMC virus, which, in the strains of mice that they used, did not produce severe diabetes.

Our studies with the D-variant of EMC virus do not support an immune mechanism (25). We have tested athymic nude mice, thymectomized mice, and chemically immunosuppressed mice and have found them no less susceptible to EMC virus-induced diabetes than normal animals (Table 3). Furthermore, immunosuppression by anti-lymphocyte serum (ALS) did not prevent the induction of EMC virus-induced diabetes. Passive transfer of lymphocytes from spleen or peripheral blood of highly diabetic SJL/J mice into normal SJL/J mice failed to produce diabetes. In crosses of susceptible and

TABLE 3: Induction of Diabetes in Nude Mice*

Strains	Virus Infected	Glucose Index	Diabetes (%)
CD-1 nu/nu	+	339 ± 109	86
	-	135 ± 14	0
CD-1 +/nu	+	249 ± 90	71
	-	142 ± 17	0
NIH Swiss nu/nu	+	211 ± 90	45
	-	132 ± 10	0
NIH Swiss +/nu	+	200 ± 72	33
	-	125 ± 8	0

* Mice with a calculated glucose index greater than 5 SD above the mean of uninfected controls were considered diabetic.

TABLE 4: Relationship of Diabetes and H-2 Haplotypes in Crosses of SJL/J and C57BL/6J Mice[a]

Strain	H-2 Genotype[b]	Number of Animals Tested	Number of Diabetic Animals[c]	Diabetes (%)
SJL/J	s/s	77	75	97
C57BL/6J	b/b	50	0	0
F1	b/s	45	5	11
F1 x SJL/J	s/s	65	33	51
	b/s	55	29	53
F1 x C57BL/6J	b/b	52	2	4
	b/s	37	2	5

(a) Male mice were inoculated i.p. with 1 x 10^5 PFU of EMC-D virus. Control mice were inoculated with an equal volume of Eagle's minimal essential medium with 5% fetal bovine serum (the vehicle used for the virus).

(b) H-2 genotypes were determined by hemagglutination.

(c) An infected animal was considered diabetic if its glucose index exceeded the mean of uninfected controls by 3 SD.

resistant mice, the inheritance of susceptibility did not show an association with H-2 haplotype (Table 4). Recently, Vialettes and colleagues failed to find evidence for a thymic-dependent, cell-mediated immune response in EMC (M-variant) virus-induced diabetes (26). Studies of the pathogenesis of EMC virus-induced diabetes mellitus also argue against an immune component. As do other picornaviruses, EMC virus rapidly infects and lyses host cells. In the case of EMC virus infection of pancreatic beta cells, evidence of this lysis can be seen within 24 to 48 hours. The virus destroys large numbers of beta cells and many of the surviving cells contain viral antigens in the cytoplasm, as determined by immunofluorescence. In this early phase of the infection, there is little or no evidence of an inflammatory response. There is also no indication that EMC virus modifies the host cell surface by inserting viral antigens or that the virus triggers an autoimmune response (e.g., no evidence of islet cell cytoplasmic antibody or islet cell surface antibody). In the case of EMC-D virus-induced diabetes, at least in the strains of mice used, the contribution of the immune response to the pathogenesis of the disease appears, at the most, to be very minor. In contrast to the negative results of studies on autoimmune mechanisms, the severity of diabetes correlated closely with the degree of beta cell damage. The degree of beta cell damage also correlated closely with the degree of virus multiplication in the beta cells. Furthermore, EMC virus-induced diabetes in mice can be prevented by a live attenuated vaccine (27). Thus it is concluded that the D variant of EMC virus can infect, multiply and directly destroy pancreatic beta cells in genetically susceptible mice independently of an autoimmune process.

5. Long-Term Complications of EMC Virus-Induced Diabetes: The separation of the M-variant of EMC virus into the D- and B-variants has made it possible to study some of the long-term complications of diabetes (28). The D variant, in the absence of the B variant, produces far more severe and prolonged diabetes than the original M variant of EMC virus. The kidney of mice that had been diabetic for 6 months showed by light microscopy both diffuse and nodular type of glomerulosclerosis (Fig. 7), and electron microscopy revealed a two- and four-fold increase in the thickness of the glomerular basement membrane. These findings are typical of those seen in humans with the Kimmelstiel-Wilson type of diabetic

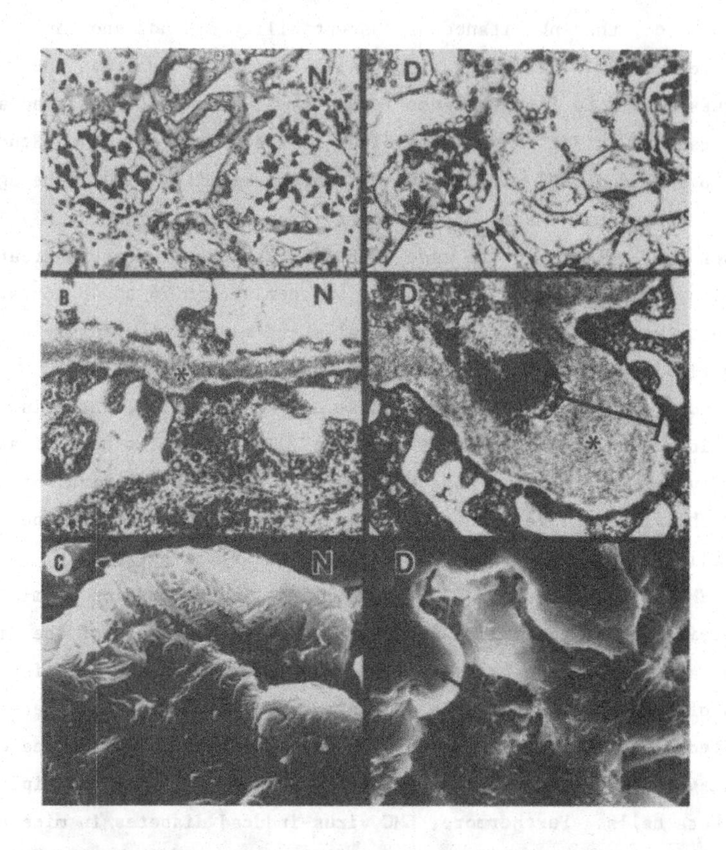

Fig. 7. Kidney sections from uninfected controls and EMC-infected mice that were diabetic for 6 months. Light microscopy of (a) control mouse showing normal glomeruli, tubules and Bowman's capsule (double arrows). (b) diabetic mouse (6 months duration) showing prominent nodular glomerulosclerosis (long arros) and thickening of Bowman's capsule (double arrows) (periodic acid-Schiff. X 145). Transmission electron micrographs of kidney from a diabetic mouse (d) showing marked thickening of the peripheral glomerular capillary basement membrane (I——I) compared with an uninfected control (c) (X 15,750). Scanning electron micrograph of kidney from diabetic mouse showing (f) focal effacement of the normal glomerular epithelial cells with loss of foot processes (arrow) compared with (e) the usual surface pattern of epithelial cells having abundant arborizing foot processes (double arrow) in normal kidney (X 3,700). Sections were prepared by standard methods and examined using a Philips 400 electron microscope or a JEOLSM35 scanning electron microscope.

glomerulosclerosis. In addition to the glomerular changes, the diabetic animals showed some of the same type of early ocular changes found in retinal vessels (e.g., a decrease in pericytes) of patients with diabetes mellitus (Fig. 8). In addition, there was a four- to six-fold increase in mortality of the highly diabetic animals as compared to control animals. Furthermore, decreased bone formation and mineralization which were found

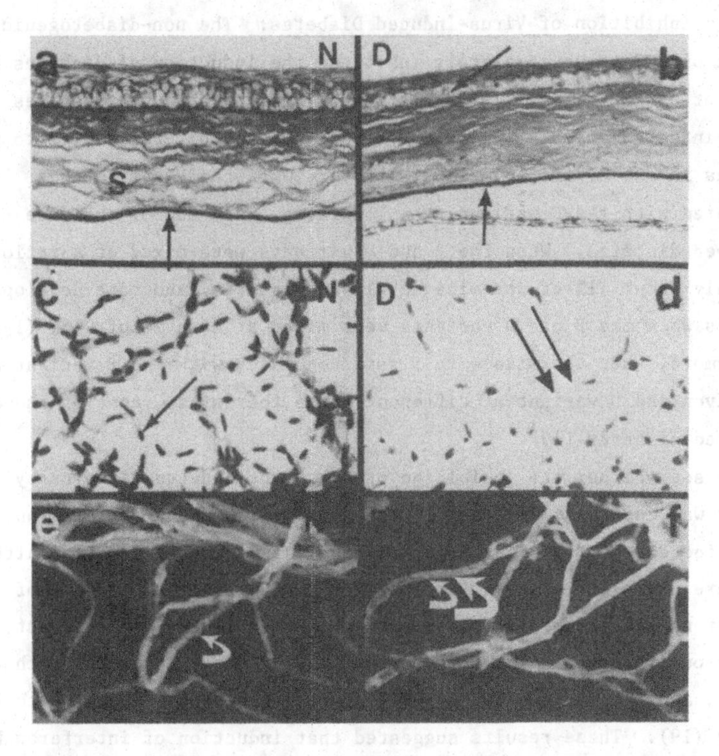

Fig. 8. The corneal epithelium of uninfected mice (a) show the usual stratification (cross) and glycogen content (periodic acid-Schiff; x 236). Cornea from mice that were diabetic for 6 months (b) show irregular stratification and moderate oedema of the epithelium, particularly of the basal layer (long arrow) (periodic acid-Schiff; x 230). The stroma (S) and Descemet's membrane (short arrow) are normal. (c) Micrograph of retinal capillaries obtained by trypsin digestion from uninfected mice, showing the normal pericyte distribution (arrow). (d) A significant decrease in the number of pericytes is observed in retinal capillaries from the diabetic mice. (e) Scanning electron micrograph of retinal capillaries obtained by trypsin digestion from uninfected mice, showing the normal pericyte distribution (arrows). (f) A moderate decrease in the number of pericytes is observed in the trypsin digest of retinal capillaries from mice that were diabetic for 6 months (x 308).

in diabetic patients, were also seen in diabetic mice at 30-180 days after infection with EMC-D virus (29). These endocondral bone changes are not due to the virus-induced tissue damage, but to the persistent metabolic alterations. Thus, the animal model is valid in the sense that virus can produce both the early metabolic changes and at least some of the long-term complications of diabetes.

6. Inhibition of Virus-Induced Diabetes: The non-diabetogenic B variant of EMC virus completely inhibited the induction of diabetes by the D variant of EMC virus: 10^4 PFU of the D variant produced diabetes in 95% of the infected animals, while the B variant, at 10^4 PFU, failed to produce diabetes in any of the animals. Moreover, when the B variant was inoculated with the D variant at a 1:1 ratio, only 60% of the mice developed diabetes. When the B and D variants were mixed at a ratio of 9:1, only about 11% of the mice developed diabetes, and none developed diabetes when the B and D variants were mixed at a ratio of 99:1 (19). Furthermore, when the mice were first inoculated with the B variant and then given the D variant at different times thereafter, none of the animals developed diabetes (19).

To see whether the inhibition of D variant-induced diabetes by the B variant was due to differences in the capacity to induce interferon, mice were infected with the D or B variant and, at various times thereafter, sera were drawn and assayed for interferon. Substantial amounts of interferon were induced by the B variant within 12 hr. In contrast, peak interferon titers were not reached until 30 hr after infection with the D variant, and the maximum titers were approximately 30% of that with the B variant (19). These results suggested that induction of interferon by the B variant may inhibit replication of the D variant at an early stage of infection. Because interferon disappeared from the circulation within 4 days after inoculation of the B variant, protection observed after 4 days after infection is most likely a result of the presence of cross-reacting neutralizing antibody. Thus, interferon (early) and antibody (later), acting in combination, appear to contribute to the B-variant-induced inhibition of diabetes in mice infected with the D variant of EMC virus.

Recent studies have shown that virus-induced diabetes can be prevented by a live attenuated vaccine (27). Mice were immunized with the non-diabetogenic B variant of EMC virus and challenged 30, 43 or 90 days later with the diabetogenic D variant of EMC virus. Diabetes did not develop in any of the immunized mice, but it did develop in approximately 80% of the unimmunized controls (Table 5). Before the challenge, serum samples were drawn and saved for measuring the titer of neutralizing antibody to the D variant of EMC virus. Neutralizing antibody titers were over 1000 at 30 days after immunization and close to 500 at 90 days. These studies show that, at least in mice, virus-induced diabetes can be prevented by a live attenuated vaccine (27). In addition to prevention of diabetes by live-attenuated vaccine, EMC virus-induced diabetes can also be prevented by the repeated administration of interferon or an interferon inducer (e.g., poly I:C) (30).

TABLE 5: Prevention of Virus-Induced Diabetes by Immunization with a Nondiabetogenic Variant*

	Infected					
	Unimmunized		Immunized		Uninfected	
Days After Immunization	Glucose Index	Diabetes (%)	Glucose Index	Diabetes (%)	Glucose Index	Diabetes (%)
30	339 ± 78	87	162 ± 12	0	155 ± 14	0
43	378 ± 84	78	172 ± 13	0	158 ± 13	0
90	307 ± 89	75	168 ± 15	0	161 ± 16	0

* Any mouse with a glucose index greater than 240 mg/dL, which was 5 SD above the mean of uninfected mice, was scored as diabetic. Each group contained 10 to 20 mice.

B. Mengovirus: Antigenically, the D variant of encephalomyocarditis (EMC) virus and the 2T variant of Mengovirus cannot be distinguished by hyperimmune sera; however, these two viruses are quite different in their tissue tropisms and mortality in mice (31). The D variant of EMC virus induces diabetes mellitus in strains of susceptible mice, but does not produce neuropathology or kill mice (32). In contrast, Mengovirus 2T produces severe brain cell damage and high mortality (31, 32). By nucleic acid hybridization, these two viruses differ by about 20 percent; moreover, binding studies suggest that they recognize different viral receptors on the same cell (31).

The plaque-purified Mengovirus 2T infects and destroys pancreatic beta cells as demonstrated by immunofluorescence and histopathological examination (Fig. 6 E and F). However, it has been difficult to study the tropism of the virus for beta cells because of the rapid onset of paralysis and death after infection. Therefore, the capacity of this virus to infect the beta cells and produce diabetes has to be studied early in the course of infection. It is interesting to note that the spectrum of host susceptibility to Mengovirus-induced diabetes is strikingly different from that produced by the D variant of EMC virus. Both viruses produced diabetes in SJL/J mice, but only Mengovirus produced abnormal glucose tolerance tests in the strains of mice which are resistant to EMC-induced diabetes such as C57BL/6J, CBA/J, C3H/J, CE/J and AKR/J (32). Immunofluorescent studies revealed that Mengovirus infects pancreatic beta cells from those mice, but EMC virus did not. Moreover, examination of islets of Langerhans from Mengovirus-infected mice revealed marked necrosis, severe inflammatory infiltration, and a decrease in the insulin content of the pancreas. To determine whether destruction of islet cells correlated with viral replication, islets from C3H/J mice infected with either Mengovirus or EMC virus were assayed for infectious virus. At one and two days after infection, a significantly higher titer (10 to 100 fold) of the virus was observed in the mice infected with Mengovirus. These studies showed that strain differences in the induction of diabetes by EMC virus and Mengovirus could be due to the degree of virus replication in pancreatic beta cells (32).

The precise mechanism by which Mengovirus infects pancreatic beta cells in the strains of mice resistant to EMC virus is not known. One of the many possibilities is that Mengovirus and EMC virus are distinct viruses that bind to different receptors on the beta cell surface. In different mouse strains, there may be quantitative differences in the expression of those receptors. Evidence in support of such a mechanism has been reported by Morishima et. al., who found that the rate of binding of Mengovirus to neuronal cell lines was five to ten times greater than the rate of EMC virus binding (31). Furthermore, receptor saturation experiments showed that unlabeled Mengovirus and EMC virus effectively blocked the binding of labeled homologous, but not heterologous virus, suggesting that these two viruses have different receptors on the cell surface. Thus, receptors specific for Mengovirus may be broadly expressed on the beta cells of mice, while those for EMC virus may be restricted to a few strains.

Regarding pathogenic mechanism, mengovirus can also infect, replicate and directly destroy pancreatic beta cells independently of autoimmune responses. We have found neither evidence of disturbances in T-cell subpopulations nor the production of autoantibodies against pancreatic islet cells in mengovirus-induced diabetic animals. However, it is difficult to exclude the possibility that neurologically regulated hormones may contribute to some extent to the abnormalities in glucose homeostasis.

C. Coxsackie B Viruses: Earlier studies showed that Coxsackievirus B4 did not produce diabetes when inoculated into mice (33). However, by repeatedly passaging Coxsackievirus B4 in murine-enriched pancreatic beta cell cultures (approximately 70% beta cells), it has been possible to enhance the diabetogenic capacity of this virus (34). However, it is very difficult to get a pure beta cell tropic virus since the cells in primary cultures are a mixture of several different cell types (34, 35). Coxsackie virus B4 that had been passaged 14 times in cultures enriched for pancreatic beta cells prepared from SJL/J mice can infect and destroy the pancreatic beta cells in certain strain of mice (Fig. 6 C and D). The destruction of beta cells resulted in a decrease in insulin content of the pancreas. This in turn led to hypoinsulinemia and the subsequent development of hyperglycemia. The reduction of immunoreactive insulin correlated inversely with the elevation of blood glucose. This, the destruction of beta cells by the Coxsackievirus B4 appears to be responsible for the development of diabetes (34, 35). As in the case of the M-variant of EMC virus, the degree of beta cell damage is, in all probability, responsible for the observed differences in the metabolic response of individual animals. In the majority of animals, hyperglycemia is transient. This may very well be due to the fact that a sufficient number of beta cells are left intact after the infection so that proliferation and/or hypertrophy of these cells result in metabolic compensation. During the acute phase of the infection, viral antigens were found in the islet of Langerhans (Fig. 6C).

The capacity of Coxsackievirus B4 to induce diabetes is influenced by the genetic background of the host. As in the case of EMC virus, only certain inbred strains of mice developed diabetes when exposed to Coxsackievirus B4, and male mice developed more severe diabetes than female mice. Moreover, the strains of mice known to be susceptible to EMC-induced diabetes were also found to be susceptible to Coxsackievirus B4-induced diabetes. Similarly, the strains of mice that were resistant to EMC-induced diabetes did not develop diabetes when exposed to Coxsackievirus B4. The only exception thus far appears to be DBA/1J and DBA/2J mice, which developed diabetes when infected with EMC virus but were resistant to the disease when exposed to Coxsackievirus B4.

II. Virus-Induced Autoimmunity and Transient and Mild Hyperglycemia.

In the previous section, the infection of beta cells with EMC, Mengo and Coxsackie B viruses, and development of diabetes in mice were discussed. In contrast to EMC, Mengo and Coxsackie B viruses, reovirus type I seem to be somehow associated with autoimmunity in the pathogenesis of a diabetes-like syndrome in mice.

A. Reovirus Type I: Since reoviruses produce a variety of lesions in newborn mice, we passaged reovirus type 3 in cultured pancreatic beta cells to see whether the virus can be adapted to the beta cells. When the beta cell-passaged virus was infected into suckling SJL/J male mice, some of the infected animals showed an abnormal response in glucose tolerance tests 10 days after infection (36). By immunofluorescence, specific viral antigens were found in some beta cells as well as in acinar cells. By electron microscopy, viral particles were detected in the cytoplasm of some beta cells. Surviving animals remained mildly hyperglycemic for about three weeks, and then returned to normal.

More recently, mice infected with reovirus type 1, which was passaged in pancreatic beta cell cultures, developed transient diabetes and a runting syndrome (12). The runting syndrome consisted of retarded growth, oily hair, alopecia and steatorrhea. Inflammatory cells and viral antigens as well as virus particles were found in the islets of Langerhans (alpha, beta and delta cells) (Fig. 2B) as well as in the anterior pituitary (growth hormone-producing cells). Examination of sera from infected mice revealed autoantibodies that reacted with cytoplasmic antigens in the islets of Langerhans, anterior pituitary, and gastric mucosa of uninfected mice. To rule out the role of these autoantibodies in the pathogenesis of reovirus-induced diabetes, infected SJL and NFS mice were treated with different immunosuppressive drugs. The administration of either anti-lymphocyte serum, anti-thymocyte serum, or cyclophosphamide reduced or prevented the development of reovirus-induced diabetes (37). In addition, virus-infected immunosuppressed mice gained weight at almost the same rate as uninfected controls, and mortality was greatly decreased. Thus, Onodera et. al., concluded that autoimmunity does play a role in the pathogenesis of reovirus-induced diabetes (12).

Precisely how reovirus infection triggers the development of auto-antibodies is still unclear, but viruses have often been suspected as a cause of autoimmune disease. In contrast to reovirus type 1, reovirus type 3 does not induce autoantibodies in mice and does not infect the pituitary. The critical difference between reovirus type 1 and type 3 seems to reside at the level of the sigma-1 polypeptide responsible for virus tropism. Therefore, it is speculated that a single viral molecule appears to control pituitary infection and autoantibody production. In view of the histopathological changes seen in lymphoid organs of reovirus-infected mice, it could be possible that specific subsets of lymphocytes may have receptors for the virus. Thus, both infection of hormone-producing cells and infection of virus with cells of the immune system may be required to initiate the production of autoantibody (12). The precise relationship between infection of reovirus type 1, the induction of autoantibodies and development of diabetic syndrome remains to be determined.

III. Persistent Infection of Pancreatic Beta Cells and Mild Hyperglycemia.

In the previous two sections, cytolytic infection of beta cells by EMC virus and reovirus-induced autoimmune-mediated destruction of beta cells were discussed. More recently, it was shown that lymphocytic choriomeningitis (LCM) virus produces a persistent infection in beta cells and mild glucose intolerance in mice (38).

A. Lymphocytic Choriomeningitis Virus: Oldstone and colleagues reported that lymphocytic choriomeningitis (LCM) virus persistently infects murine pancreatic islet cells (38). In their studies, viral nucleoprotein was detected predominantly in the pancreatic beta cells by double-labeled immunofluorescent antibody technique. Electron microscopy confirmed these findings by showing virions budding from the beta cells. Persistent infection was associated with the chemical evidence of diabetes including hyperglycemia and abnormal glucose tolerance. However, the virus-infected islet cells showed normal anatomy and cytomorphology. Neither beta cell destruction nor lymphocytic infiltration was routinely observed. The end result is a chemical and morphologic picture similar to that observed in early stages of adult-onset diabetes mellitus.

One of the possible mechanisms by which the virus might cause diabetes is through the establishment of an infection that might shut off the "luxury functions" of insulin-producing beta cells (38, 39). The other possibility is that the persistent infection may also result in a gradual reduction in the number of functioning beta cells since the regenerative capacity of beta cells is thought to be poor (39). The precise mechanism by which LCM-virus induces diabetes remains to be investigated. Nevertheless, this new finding is very interesting and important for studies on diabetes in humans and animals.

ACKNOWLEDGEMENTS

This work was supported by grants from the Medical Research Council of Canada (DQ336 and MA9584), and the Canadian Diabetes Association. The author is a Heritage Medical Scientist of the Alberta Heritage Foundation for Medical Research. The editing and secretarial help of Judy Crawford is greatly acknowledged.

REFERENCES

1. Yoon, J.W. and Notkins, A.L. Metab. (Suppl.) 32:37-40, 1983.
2. Notkins, A.L., Yoon, J.W., Onodera, T., Toniolo, A., and Jenson, A.B. In Prospectives in Virology 11:141-162, 1981.
3. Craighead, J.E. Progr. Med. Virol. 19:161-214, 1975.
4. Yoon, J.W. In Current Problems in Clinical Biochemistry (Eds. H. Kolb, G. Schernthauer, F.A. Fries), Bern, Stuttgart, and Vienna, Han Huber Publishers, 1983, pp. 11-37.
5. Yoon, J.W., Austin, M., Onodera, T. and Notkins, A.L. N. Engl. J. Med. 300:1173-1179, 1979.
6. Gladisch, R., Hofmann, W. and Waldherr, R. Z. Kardoil 65:837-849, 1976.
7. Champsaur, H., Bottazzo, G., Bertrams, J., Assan, R. and Bach, C. J. Pediat. 100:15-20, 1982.
8. Ginsberg-Fellner, F., Witt, M.E., Yagihasi, S., Dobersen, M.J., Taub, F., Fedun, B., McEvoy, R.C., Roman, S.H., Davies, T.F., Cooper, L.Z., Rubinstein, P. and Notkins, A.L. Diabetologia (Suppl.) 27:87-89, 1984.
9. Craighead, J.E. and McLane, M.R. Science 162:913-915, 1968.
10. Stefan, Y., Malaisse-Lague, F., Yoon, J.W., Notkins, A.L. and Orci, L. Diabetologia 15:395-401, 1978.
11. Hayashi, K., Boucher, D.W. and Notkins, A.L. Am. J. Pathol. 75:91-102, 1974.

12. Onodera, T., Toniolo, A., Ray, U.R., Jenson, A.B., Knazek, R.A. and Notkins, A.L. J. Exp. Med. 153:1457-1473, 1981.
13. Onodera, T., Yoon, J.W., Brown, K.A. and Notkins, A.L. Nature (Lond.) 274:693-695, 1978.
14. Yoon, J.W. and Notkins, A.L. J. Exp. Med. 143:1170, 1976.
15. Yoon, J.W., Lesnick, M.A., Fussganger, R. and Notkins, A.L. Nature (Lond.) 264:178-180, 1976.
16. Chairez, R., Yoon, J.W. and Notkins, A.L. Virology 85:606-611, 1978.
17. Ross, M.E., Onodera, T., Brown, K.S. and Notkins, A.L. Diabetes 25:190-196, 1976.
18. Yoon, J.W., Onodera, T. and Notkins, A.L. J. Gen. Virol. 37:225, 1977.
19. Yoon, J.W., McClintock, P.R., Onodera, T. and Notkins, A.L. J. Exp. Med. 152:878-892, 1980.
20. Ray, U., Aulakh, G., Schubert, M., McClintock, P.R., Yoon, J.W. and Notkins, A.L. J. Gen. Virol 64:947-950, 1983.
21. Yoon, J.W. In: Genetic Environmental Interaction in Diabetes Mellitus (Eds. J.S. Melish, J. Hanne, and S. Baba), Excerpta Medica, (International Congress Series No. 549), 1982, pp. 227-234.
22. Jansen, F.K., Munterfering, H. and Schmidt, W.A.K. Diabetologia 13:545-49, 1977.
23. Buschard, K., Rygaard, J. and Lund E. Acta. Pathol. Microbiol. Scand. [C] 84:299-303, 1976.
24. Buschard, K., Hastrup, N. and Rygaard, J. Diabetologia 24:42-46, 1983.
25. Yoon, J.W., McClintock, P.R., Bachurski, L.J., Longstreth, J.D. and Notkins, A.L. Diabetes 34:922-925, 1985.
26. Vialettes, B., Baume, D., Charpin, C., De Maeyer-Guignard, J. and Vague, P. J. Clin. Lab. Immunol. 10:35-40, 1983.
27. Notkins, A.L. and Yoon, J.W. N. Engl. J. Med. 306:486, 1982.
28. Yoon, J.W., Rodrigues, M.M., Currier, C. and Notkins, A.L. Nature 296:567-570, 1982.
29. Yoon, J.W. and Reddi, A.H. Am. J. Physiol. 246:C177-C179, 1984.
30. Yoon, J.W., Cha, C.Y. and Jordan, G. J. Inf. Diseases 147:155-159, 1983.
31. Morishima, T., McClintock, P.R., Aulakh, G.S., Billups, L.C., and Notkins, A.L. Virology 122:461-465, 1982.
32. Yoon, J.W., Morishima, T., McClintock, P.R., Austin, M. and Notkins, A.L. J. Virol. 50:684-690, 1984.
33. Ross, M.E., Onodera, T., Hayashi, K. and Notkins, A.L. Infect. Immun. 12:1224-1226, 1975.
34. Yoon, J.W., Onodera, T. and Notkins, A.L. J. Exp. Med. 148:1068-1080, 1978.
35. Toniolo, A., Onodera, T., Jordan, G., Yoon, J.W. and Notkins, A.L. Diabetes 31:496-499, 1982.
36. Onodera, T., Jenson, A.B., Yoon, J.W. and Notkins, A.L. Science 301:529-531, 1978.
37. Onodera, T., Ray, U.R., Melez, K.A., Suzuki, H., Toniolo, A., Notkins, A.L. Nature 297:66-68, 1982.
38. Oldstone, M.B.A., Southern, P., Rodriguez, M. and Lampert, P. Science 224:1440-1443, 1984.
39. Notkins, A.L. and Yoon, J.W. In: Concepts in Viral Pathogenesis` (Eds. A. L. Notkins and M.B.A. Oldstone), New York, Springer-Verlag, pp. 241-247, 1984.

14

THE INTERFERON SYSTEM IN ENCEPHALOMYOCARDITIS (EMC) VIRUS-INDUCED DIABETES MELLITUS

S.H. COHEN AND G.W. JORDAN

Division of Infectious & Immunological Diseases Department of Internal Medicine University of California at Davis School of Medicine Sacramento, California

ABSTRACT

Two closely related plaque variants selected from the M-strain of encephalomyocarditis (EMC) virus differ in their ability to cause murine diabetes mellitus. EMC-D infection causes diabetes in susceptible strains of mice while EMC-B does not cause diabetes but EMC-B interferes with the production of diabetes by the D variant. These characteristics are due to the greater ability of EMC-B to induce interferon in mice compared to EMC-D. Concomitantly, EMC-B is more sensitive to the action of exogenous interferon because local interferon production is amplified in interferon-primed cells infected with EMC-B. These properties are determined by the interferon-inducing particle (ifp) phenotype of the variants. EMC-B is phenotypically ifp^{+} whereas EMC-D is ifp^{-}. The ifp phenotype is a critical factor in determining the outcome of infections with the B and D variants of EMC virus.

INTRODUCTION

In 1968, Craighead and McLane showed that the inoculation of genetically susceptible strains of mice with the M-strain of encephalomyocarditis (EMC) virus results in diabetes mellitus. They showed that the virus replicates in pancreatic beta cells, and causes insulitis and hyperglycemia (1). However, not all infected animals

Becker, Y (ed), Virus Infections and Diabetes Melitus. © 1987 Martinus Nijhoff Publishing, Boston. ISBN 0-89838-970-4.

become diabetic and the proportion of diabetic mice varies with the passage history of the virus and the extent of damage to the islets of Langerhans (2,3). In 1978 Yoon et. al. plaque selected two variants of the M-strain for and against the ability to cause diabetes. The EMC-D variant causes diabetes in 100% of susceptible mice whereas EMC-B, which can infect the mice and spread to the pancreas, does not cause diabetes. Diabetes caused by EMC-D is present 3 days after infection in association with beta-cell destruction. Additional experiments with this model demonstrate that co-inoculation of mice with different proportions of the two variants results in metabolic abnormalities of varying severity because EMC-B interferes with the production of diabetes by EMC-D (4,5). Because these two variants were present together in the M-strain which is only partially diabetogenic, this model provides an opportunity to study the mechanism by which the B and D variants interact to modulate the outcome of infection.

INTERFERON SYSTEM IN VIVO

Infection of mice with EMC-B results in higher titers of circulating interferon in comparison with mice infected with EMC-D. Maximum circulating interferon titers induced by EMC-B appear earlier and are about three times higher than those following infection with EMC-D (4). Higher titers of interferon are also observed within hours in the peritoneal fluid, the site of inoculation in this model, following infection with EMC-B when compared with EMC-D (6). These differences in the interferon response of mice to the B and D variants which occur both at the site of inoculation and in the blood and are detectable after only a few hours suggest a role for the interferon system in determining the diabetic outcome of this infection.

This hypothesis is supported by several lines of evidence. First, interferon or interferon inducers prevent murine diabetes mellitus following EMC-D virus infection. Repeated injections of poly I:C or mouse fibroblast interferon are necessary and must be given both pre- and post-infection (7). Second, when SJL mice were pretreated with anti-interferon globulins to neutralize circulating interferon, the titers of EMC-B virus in pancreas, brain and heart were comparable to those attained by the D variant in tissues of untreated mice. Also, a four-fold rise in the number of pancreatic islet cells containing EMC-B viral antigens was observed, indicating that the B variant infected more cells when circulating interferon was neutralized. Third, although infection with EMC-B in untreated mice is asymptomatic, in SJL mice given anti-interferon globulins the mortality rate was 40% and of the survivors 40% became diabetic (7). Therefore, use of anti-interferon globulins demonstrated that the interferon system is one determinant of the diabetic outcome in EMC virus infection.

Other comparisons of the B and D variants have not revealed differences that could explain the difference in diabetic outcome. The B and D variants are antigenically similar and are indistinguishable by reciprocal neutralization tests. Neutralizing antibody reaches peak titers by the seventh day after infection of SJL mice with EMC virus. Comparison of mice infected with the EMC-B or EMC-D variants revealed no difference in the time of appearance or titer of neutralizing antibodies (4). Because insulin deficiency and hyperglycemia are present at 48 hours after infection, before the humoral immune response is detectable, the difference in outcome of infection with EMC-B and D is not related to the humoral immune system.

Cultured mouse beta-cells can be infected by a variety of nondiabetogenic EMC viruses. Similarly, beta cells from strains of mice which do not become diabetic are susceptible to infection by the M-stain of EMC virus _in vitro_ (8). The number of pancreatic islet cells containing viral antigens _in vivo_ can be increased by pretreatment of the mice with anti-interferon globulins as previously described (7). Therefore, the development of diabetes after EMC-D, but not EMC-B, infection does not appear to be determined by the presence or absence of cell receptors for the respective viruses.

Taken together, these results demonstrate that the induction of interferon normally limits the replication and diabetogenicity of EMC-B virus. Conversely, it would appear that the relative inability of EMC-D to induce interferon is a virulence factor.

INTERFERON INDUCTION _IN VITRO_

The evidence indicating a role for the interferon system _in vivo_ as an explanation for biological differences between B and D led to studies of the relationship of the interferon system to the growth of these variants _in vitro_. Mouse embryo cells co-infected with EMC-B and D yield reduced amounts of EMC-D. EMC-B induces more interferon than EMC-D in mouse embryo and mouse L-cells _in vitro_ which correlates with this phenomenon of interference (9). Also, murine peritoneal macrophages inoculated _in vitro_ produce more interferon following infection with EMC-B than with EMC-D (6). Thus the ability of EMC-B to induce more interferon and to interfere with the replication of EMC-D in cell culture parallels the _in vivo_ ability of EMC-B to interfere with the production of diabetes by EMC-D.

SENSITIVITY TO INTERFERON IN VITRO

EMC-B was shown to be more sensitive to inhibition by exogenous interferon than EMC-D in experiments comparing interferon-induced inhibition of virus yield and plaque reduction on mouse embryo cells and mouse L-cells (Ly), cell lines which respond to interferon inducers. This difference was most obvious at low interferon concentrations and low multiplicities of infection (9). A line of L-cells (Lsp) that responds to exogenous interferon, but not to interferon inducers, was used to study the mechanism of the different interferon sensitivity of EMC-B and EMC-D. Using this cell line, B and D were equally sensitive to exogenous interferon, indicating that the greater interferon sensitivity of EMC-B is due to its ability to induce additional (endogenous) interferon in cells pretreated with interferon (9). Interferon-treated cells produce higher titers of interferon more rapidly than untreated cells, a property called "priming" (10). The priming phenomenon was also demonstrated when Ly cells were infected with the B and D variants and observed for cytopathic effect. In untreated cells, both viruses caused cytopathic effects, but when these cells were pretreated with interferon, greater protection was seen in cells infected with EMC-B. The difference in the cell sparing effect on Ly cells was most prominent at low interferon concentrations (Fig. 1).

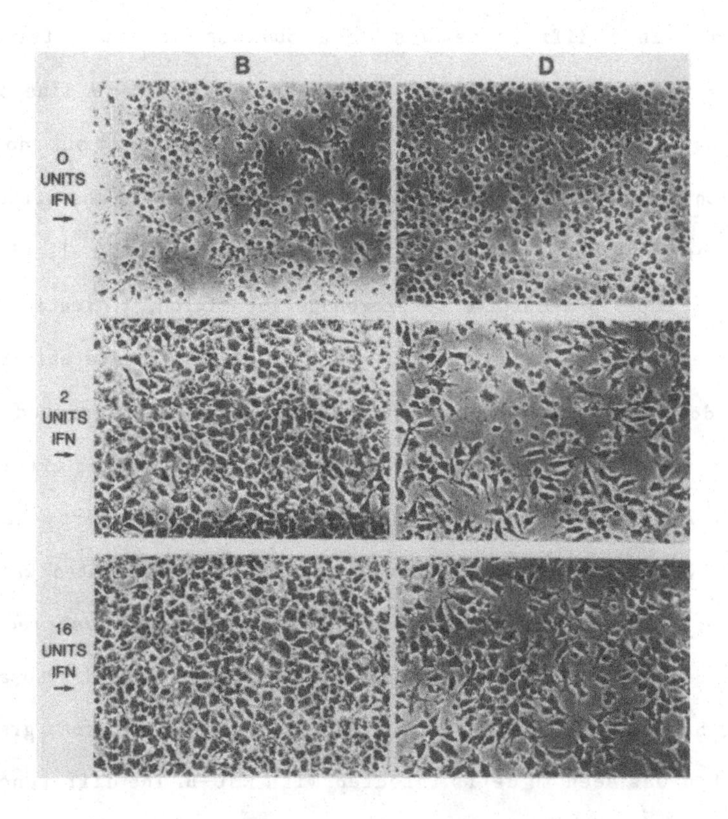

Figure 1

Mouse Ly cells treated with mouse fibroblast interferon and infected with the B or D variant of EMC virus. The cells in 60 mm petri dishes were pretreated with 2 or 16 units of interferon for 18 hours at 37°C and inoculated with virus at a multiplicity of infection (M.O.I.) of 4. After a one hour absorption period 3 ml of MEM-2 were added and cultures were incubated. Photomicrographs were taken at 24 hours post infection.

In contrast, equal protection against the cytopathic effects of B and D was observed on interferon-treated Lsp cells which do not respond to interferon inducers (9).

To further demonstrate the role of locally induced interferon, monolayers of Ly and Lsp cells were exposed to various concentrations of poly I-C for 18 hours before infection (9). L cells respond to poly I-C by becoming sensitized for augmented interferon production but no interferon is produced until a viral infection or other induction event occurs (10, 11, 12). Virus replication, as measured by plaque formation, on Ly cells after EMC infection was reduced by pretreatment with 10 µg/ml of poly I-C. The inhibition of both size and number of plaques was greater for EMC-B as compared with EMC-D. Pretreatment with poly I-C did not affect plaque number or size of either variant when Lsp cells were infected (9). These results indicate that EMC-B induces additional interferon locally which is responsible for its increased sensitivity to interferon.

The difference in interferon sensitivity of the EMC variants is similar to observations made for a mutant of mengovirus, a related picornavirus of mice. Simon et. al. isolated an interferon-sensitive (is-1) mutant of mengovirus after mutagenesis with nitrous acid (13). Marcus et.al. established that the is-1 mutant of mengovirus is interferon-sensitive because it induces endogenous interferon which further increases resistance of the infected cells. In contrast, the wild type does not have this property (14). This pair of mengovirus variants was used to study interferon induction. The is-1 variant induced high titers of interferon while the wild type induced none. Thus, two different interferon-inducing particle (ifp) phenotypes were defined when the initial multiplicity of infection was maintained using anti-mengovirus serum. The relationship between interferon yield and input multiplicity indicated that a single particle per cell induced a quantum yield of interferon while a second particle inhibited interferon production. Thus, the is-1 variant was defined as ifp$^+$. The ifp$^+$ phenotype is associated with increased sensitivity to interferon and is less cytopathic to cells because the priming effect amplifies local interferon production. Furthermore, because a second ifp$^+$ particle per cell suppresses interferon production, these phenomena are observed best at low multiplicities of infection. These phenomena are also best observed at interferon concentrations which are high enough to prime cells for interferon production but insufficiently high to completely inhibit viral growth. In contrast, the ifp$^-$ phenotype of the mengovirus wild type is associated with relative interferon insensitivity and greater cytopathology (14).

Though each ifp$^+$ particle induces a quantum amount of interferon per cell, viruses with the ifp$^+$ phenotype can interact with cells to induce interferon in two fundamentally different ways when the initial input multiplicity is maintained by antisera (15). In the type 1 response, the interferon yield increases with increasing multiplicity of infection as a greater percentage of cells are infected with one or more viral particles and induced to produce interferon. The number of cells infected increases until it plateaus; similarly the interferon yield increases and plateaus at a maximum yield. This type of multiplicity-interferon yield response is seen when aged chick embryo cells are infected with avian reovirus. The type 2 response differs in that a single ifp$^+$ particle induces a quantum amount of interferon, whereas interferon induction is suppressed in cells infected with two or more ifp$^+$ particles as has been described with the mengovirus is-1 mutant (15). The B and D variants of EMC virus are analogous to the interferon-sensitive mutant and wild-type mengovirus. EMC-B is phenotypically ifp^{+}, exhibiting a type 2 induction response on L cell monolayers on which the input multiplicity is maintained with anti-EMC serum while EMC-D induces very little interferon & is phenotypically ifp^{-} (9) (Fig. 2).

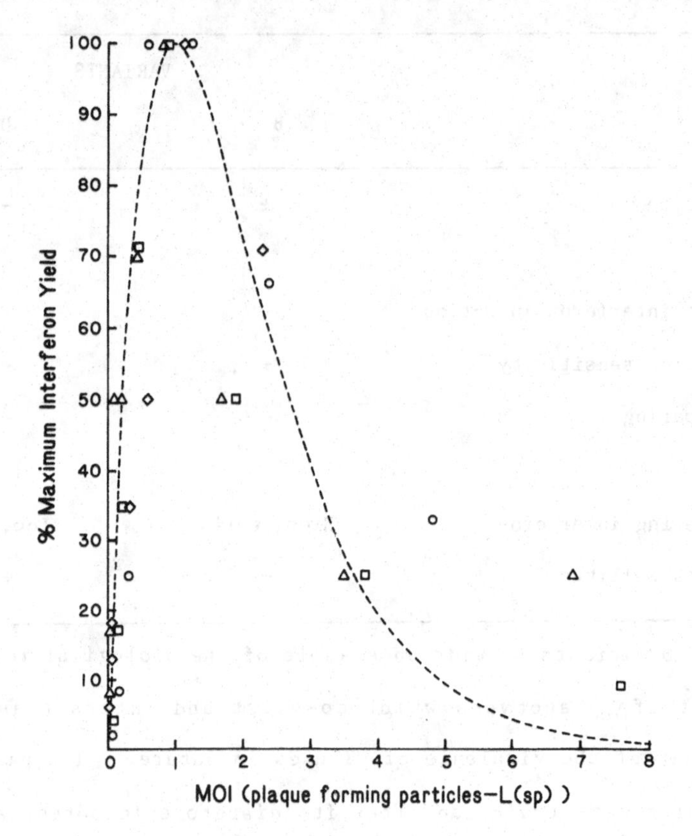

Figure 2

Dose-response curve relating multiplicity of EMC-B, maintained with anti-EMC serum, to the interferon yield of mouse Ly cells at 18 hours. The abscissa represents the multiplicity of plaque-forming particles (PFP) per cell as assayed on mouse Lsp cells. The dashed curve represents the Poisson distribution (shown as the per cent of the maximum probability) for the class of cells infected with only one PFP (r=1) where m=MOI and $P=e^{-m}m^{r}/r!$. The symbols represent the mean of two samples from four different experiments. Maximum interferon units produced were ◇=240 □=240 △=480 and 0=960. (Reprinted with permission from Ref. 9).

THE BIOLOGICAL ROLE OF THE IFP PHENOTYPE

The B and D variants of EMC virus were selected from the M-strain based only on their ability to cause diabetes in mice (14). Concomitantly, different ifp phenotypes were obtained which have been shown to control the diabetic outcome in this model (9) (Table 1).

Table 1
Biological Properties Associated with the ifp Phenotype of EMC Virus

	VARIANTS	
	B	D
ifp Phenotype	+	-
in vitro		
One hit interferon induction	+	-
Interferon sensitivity	+	-
cell sparing	+	-
SJL mice		
Circulating interferon	High, early	Low, late
Diabetes Mellitus	-	+

The B and D variants provide an example of the biological role of different ifp phenotypes which co-exist and interact as one determinant of the virulence of viruses in nature. The passage history of the M-strain can alter its diabetogenic potential to resemble co-infection with different proportions of the B and D variants (4, 16, 17). Passage of EMC-D in different cell lines can alter its diabetogenic potential. (18) It is possible that ifp$^-$ could give rise to ifp$^+$ variants as occurred with the mengovirus wild type and is-1 mutant. Thus, the ifp phenotype must be considered along with other viral properties such as temperature sensitivity, defective interfering particles and cell receptor interactions for a complete understanding of viral pathogenesis.

CONCLUSION

The interferon system plays a major role in EMC virus-induced diabetes mellitus. The in vitro phenomena related to the ifp phenotype parallel the in vivo findings in this model infection. Therefore, we propose the following mechanism by which the ifp phenotype acts in vivo to alter the outcome of EMC viral infection. The amount of interferon produced by the peritoneal macrophages after intraperitoneal inoculation of EMC virus depends on the ifp phenotype or mixture of ifp phenotypes of the infecting strain. Relatively higher circulating interferon titers occur within a few hours of EMC-B virus infection. Circulating interferon reaches the target organs and engenders an antiviral effect. More importantly, this first exposure to interferon primes the cells of the target organ, in this case islet cells of the pancreas, to rapidly produce increased amounts of endogenous interferon locally upon exposure to the infecting virus. Once the virus reaches this local environment, the ifp phenotype is a critical factor in determining the outcome. Therefore, EMC-B will be contained by the augmented antiviral activity of the interferon-primed pancreatic cells while EMC-D encounters minimally primed pancreatic cells less able to halt tissue destruction and the resultant diabetes mellitus.

REFERENCES

1. Craighead JE, McLane MF. Science 162:913-914,1968.
2. Ross ME, Onodera T, Hayashi K, Notkins AL. Infect Imm 12:1224-1226,1975.
3. Iwo K, Bellomo SC, Mukai N, Craighead JE. Diabetologia 25:39-44,1983.

4. Yoon J, McClintock PR, Onodera T, Notkins AL. J Exp Med 152:878-982,1980.
5. Gould CL, Trombley ML, Bigley NJ, McMannama, Giron DJ. Proc Soc Exp Biol Med 175:449-453,1984.
6. Cohen SH, Acevedo V, Jordan GW. Twenty-third Interscience Conference on Antimicrobial Agents and Chemotherapy October 24-26, 1983 Las Vegas, NV Abstract No 391.
7. Yoon JW, Cha CY, Jordan GW. J Infec Dis 147:155-159,1983.
8. Wilson GL, D'Andrea BJ, Bellomo SC, Craighead JE. Nature 285:112-113,1980.
9. Cohen SH, Bolton V, Jordan GW. Infect Imm 42:605-611,1983.
10. Stewart WE II. In: The interferon system. 2nd ed. Wien, New York: Springer-Verlag 1981 pp. 223-256.
11. Schafer TW, Lockart RZ. Nature 226:449-450,1970.
12. Vengris VE, Stollar BD, Pitha PM. Virology 65:410-417,1975.
13. Simon EH, Kung S, Koh TT, Brandman P. Virology 69:727-736,1976.
14. Marcus PI, Guidon PT, Sekellick ML. J Interferon Res 1:601-611,1981.
15. Marcus P. J Interferon Res 2:511-518,1982.
16. Giron DJ, Cohen SJ, Lyons SP, Wharton CH, Cerutis DR. Proc Soc Exp Biol Med 173:328-331,1983.
17. Yoon JW, Onodera T, Notkins AL. J Gen Virol 37:225-232,1977.
18. Giron DJ, Patterson RR, Lyons SP. J Interferon Res 2:371-376,1982.

Summary

15

VIRAL MECHANISMS LEADING TO DIABETES MELLITUS - A SUMMARY

Y. BECKER

Department of Molecular Virology, Faculty of Medicine, The Hebrew University, Jerusalem, Israel

ABSTRACT

The nature of the viral mechanisms leading to destruction of pancreatic beta cells or cessation of insulin production is still not fully understood. A link between the class 1 hypervariable region (HVR) and insulin-dependent diabetes mellitus (IDDM) was noted in which autoantibodies to beta cell membrane antigens are synthesized against specific HLA types (DR-3 and DR-4) and lead to destruction of the insulin-producing beta cells. Coxsackie B4, infectious hepatitis and congenital rubella viruses find receptors on the beta cells that allow for virus infection and cell destruction. Mimicry between virus antigens and beta cell membrane antigens leads to the appearance of islet cell antibodies (ICA), as in the case of mumps virus infections. Members of the herpesvirus group infect beta cells, causing cessation of insulin production and the production of antibodies to viral antigens that crossreact with insulin (insulin autoantibodies), as occurs in chickenpox.

Helmke (1) indicates that "reports describing a remarkable coincidence of a virus infection preceding the manifestation of diabetes mellitus date back to the previous century". He quotes as the earliest report in respect to mumps infection the Norwegian physician, J. Stang who, in 1864, reported a patient developing diabetes shortly after mumps infection. These observations, and many others which followed (1), revealed the connection between infection of man with viruses such as Coxsackie B4, reovirus, rubella, mumps and herpesviruses and the development of diabetes mellitus (2-8). The chapters in the present

Becker, Y (ed), Virus Infections and Diabetes Melitus. ISBN 0-89838-970-4.

volume deal with these studies. This summary is intended to describe possible mechanisms by which viruses affect the beta cells and lead to the development of diabetes mellitus. It is based on our knowledge of the insulin gene and its hypervariable region (HVR; 9).

Two forms of the diabetes mellitus syndrome are known: IDDM - insulin-dependent diabetes mellitus (type 1), and NIDDM - noninsulin-dependent diabetes mellitus (type 2). HVR and IDDM can be described as follows*:

1. HVR, located upstream of the 5' end of the human insulin gene, is associated with IDDM. In Caucasians with IDDM, the frequencies of HVR class 1 alleles and the homozygous class 1 genotype were significantly higher than in non-diabetic or NIDDM groups.
2. IDDM is associated with class II major histocompatibility complex (MHC) antigens HLA-DR3 and DR4.
3. Absolute insulin deficiency is due to destruction of beta cells and not abnormality in the expression of the insulin gene.
4. The diabetogenic locus (HVR) near the insulin gene might contain a gene encoding a beta-cell-specific autoantigen that may determine the susceptibility of the beta cell to virus infections.

Environmental factors, in addition to a genetic predisposition to IDDM, lead to the full expression of the disease. Human viruses with a predilection for pancreatic beta cells, were suspected of being one of the environmental factors that can cause IDDM. Yoon *et al.* (10) reported in 1979 on a virus infection in a child who died of diabetic ketoacidosis. The

---

* Data from Xiang, K. *et al.* (9).

data presented in this volume on the involvement of members of several virus families in diabetes are summarized in Table 1. Viruses can cause IDDM by different mechanisms that involve direct destruction of the beta cells by infection (Table 1, B1), and possible mimicry between viral antigens and beta-cell surface antigens, leading to appearance of islet cell antibodies (ICA) (1, 11) (Table 1, B2). Varicella virus, a member of the herpesvirus family, was found to cause the appearance of antibodies to insulin (insulin autoantibodies) in 81% of chickenpox patients studied (11). In addition, it was reported by Kurata *et al.* (12) that a generalized HSV infection led to the destruction of beta cells, resulting in diabetes (Table 2B. 3 and 4).

In Table 1, B6, two additional possibilities were included as speculation: a) virus-induced destruction of central nervous system cells producing the hormone vasopressin, which is responsible for stimulating beta cells to produce insulin; and b) mimicry between an infecting virus and insulin receptor polypeptide, that could result in receptors nonresponsive to insulin. Further studies might clarify if such mechanisms of virus-induced damage play a role in diabetes. Venezuelan equine encephalitis (VEE) virus (16) was reported in several studies to lead to a reduction of insulin in monkeys. Nevertheless, VEE epidemics in Venezuela and Colombia did not increase the incidence of diabetes.

The HVR situated in chromosome 11 at position 5' to the insulin gene constitutes the 14-nucleotide sequence ACAGGGGTGTGGGG repeating numerous times (Table 2; 9). Studies on herpes simplex virus type 1 (HSV-1) (17-19) revealed that repeat sequences are present in the viral genome at a 3' position to two immediate-early (IE) genes: IE1 (18) and IE5 (19). The IE1 gene is followed by eight repeats of 16 nucleotides and IE5 by nine repeats of 15 nucleotides (Table 2). It is of interest that the insulin

Table 1. Viral infections leading to damage of insulin-producing beta cells or cessation of insulin production

Type of mechanism	Viruses
A. Non-viral, genetically controlled destruction of beta cells by autoantibodies	Genetic association with class 1 HVR and HLA DR-3 and DR-4. The diabetic locus near the insulin gene might also determine the susceptibility of the beta cells to virus infections (9)
B. Virus-induced damage to beta cells:	
1. Destruction of beta cells due to virus infection and replication. Specific predilection for beta cells due to presence of cellular glycoprotein(s)serving as virus receptors on cell surface	Coxsackie B4 (13) Infectious hepatitis (14) Congenital rubella (15)
2. Antibodies to viral antigens on beta cell membrane mimic islet-cell autoantibodies (ICA) in destruction of beta cells (mimicry)	Persistent mumps virus infection (epidemic 1977-1981) (1, 11)
3. Virus infection of beta cells lead to nuclear inclusion bodies and cessation of insulin granule production	Herpesviruses: HSV-1, varicella zoster, cytomegalovirus (12)
4. Antibodies to virus antigens resembling insulin autoantibodies (IAA) react with insulin and cause insulin depletion	Varicella virus in chickenpox (12)
5. Persistent virus infections	Mumps, congenital rubella (1, 15)

6. Possible additional mechanisms (speculation):
   a. Virus infections leading to the suppression of the hormone vasopressin which stimulates pancreatic beta cells to produce insulin
   b. Virus antigen mimicry of insulin receptors

gene HVR repeat shares the sequence AGGGGTG with the repeat sequence of the HSV-1 IE1 gene and an almost identical sequence with the repeat sequence of HSV-1 IE5 (Table 2). These two viral genes are expressed in HSV-1 infected cells early in the replicative cycle. It is possible that the repeat sequences are recognized by the host cell, subsequently leading to the transcription of the viral IE genes. Repeat sequences can be found near the Epstein-Barr herpesvirus gene which codes for the EB nuclear antigen-1 (EBNA-1) and 5' to the T antigen gene in SV40 (Table 2). Hayward *et al.* (17) indicated that a 400-repeat of nine nucleotides GAGCTGGGG is present in the immunoglobulin switch region and in introns of a number of cellular genes, such as Z-globin, γ-globin and Ig variable region (Table 2).

It is possible that the repeat sequences at or near specific genes in the cellular and viral genomes serve as recognition sites to direct the cellular transcription machinery to genes near the repeat sequences, as described for HSV-1. It is, therefore, possible to speculate that viruses infecting beta cells might have the ability to interfere with the function of the HVR 5' to the insulin gene and thus interfere with the expression of the insulin gene and cause IDDM in infected individuals. Alternatively, Xiang *et al.* (9) suggest that HVR may include a gene encoding a beta-cell-specific autoantigen which determines the sensitivity of beta cells to viral infection or its response to such infection". The predilection of Coxsackie B4 virus to beta cells may suggest the presence of a specific receptor for this virus on beta cells. In addition, the appearance of ICA antibodies after persistent mumps virus infection (1) may indicate that a viral protein shares an antigenic domain with a membrane protein specific to beta cells, and that the antiviral antibodies act as autoantibodies to beta cells and cause cell destruction and IDDM. It should be noted that mumps

Table 2 Comparison of repeat sequences in the hypervariable region of the insulin gene with repeat sequences in other cellular and viral genes

Cellular and viral genes	No. of repeats	Repeat sequence	Ref.
Insulin gene HVR		ACAGGGGTGTGGGG	9
Class 1	40x		
Class 2	95x		
Class 3	170x		
Viruses			17
HSV-1 3' to IE1	8x	TGAGGGTGCGTCGGGG	18
HSV-1 3' to IE5	9x	GGTGAGGGGTGGGTG	19
EBV BamHIK coding for EBNA-1	80x	GGGGCAGGA	17
EBV BamHIY	14x	TGGTGGGGG	17
SV40 5' toT antigen	3x	GGGCGGAGTTAGGGGCGGA	17
Mammalian cells: DNA			
Ig switch region	400x	GAGCTGGGG	17
Z-globin intron	70x	CGGGG	17
Z-globin intron	40x	ACAGTGGGG AGGGG	17
Introns of γ-globin, actin, γ-interferon, Ig variable region, 3' and 5' flanking to α-globin.	6-180x	CA	17

virus is implicated as having other effects in humans, such as brain complications and heart damage (20).

Studies in the last decade have established the importance of virus infections as possible causative agents triggering the development of diabetes mellitus type 1 (IDDM). Future studies might shed light on the exact mechanisms of virus-induced IDDM, and prevention of virus infection leading to IDDM would reduce the incidence of diabetes resulting from virus infections. Effective antiviral drugs such as acyclovir (acycloguanosine), which can be used for the treatment of generalized HSV-1 in neonates and children (12), may, in time, extirpate the disease and thus prevent deaths due to IDDM, as reported by Kurata *et al.* (12). Further studies on HVR and the proinsulin gene, and on the properties of virus-coded proteins and regulatory sequences in viral genomes, are needed for the elucidation of the role of viruses in diabetes mellitus in man.

ACKNOWLEDGMENTS

Studies on herpes simplex virus type 1 in our laboratory are supported, in part, by grants from the United States-Israel Binational Science Foundation and the Foundation for the Study of Molecular Virology and Cell Biology, Phoenix, Arizona, USA.

REFERENCES

1. Helmke, K. Chapter 7, this volume.
2. Maugh, T.H. Science 188:347-351, 1975.
3. Onodera, T., Jenson, A.B., Yoon, J.W. and Notkins, A.L. Science 201:529-531, 1978.
4. Onodera, T., Yoon, J.W., Brown, K.D. and Notkins, A.L. Nature 274:693-696, 1978.
5. Onodera, T., Toniolo, A., Ray, U.R., Jenson, A.B., Krazek, R.A. and Notkins, A.L. J. Exp. Med. 153:1457-1473, 1981.
6. Prince, G., Jenson, A.B., Billups, L. and Notkins, A.L. Nature 271:158-161, 1978.
7. Helmke, K., Otten, A. and Willenis, W. Lancet II:211-212, 1980.
8. Ginsberg-Fellner, F., Klein, E., Dobersen, M., Jenson, A.B., Rayfield, E., Notkins, A.L., Rubinstein, P. and Cooper, L.Z. Pediat. Res. 14:572. 1980.
9. Xiang, K., Sanz, N., Karam, J.H. and Bell, G.I. Chapter 1, this volume.
10. Yoon, J. W., Austin, M., Onodera, T. and Notkins, A.L. N. Engl. J. Med. 300:1173-1179, 1979.
11. Bodansky, H.J., Dean, B.M., Bottazzo, G.F., Grant, P.J., McNally, J., Hambling, M.H. and Wales, J.K. Lancet II:1351-1353, 1986.
12. Kurata, T., Sata, T., Iwasaki, T., Onodera, T. and Toniolo, A. Chapter 8, this volume.
13. Cook, S.S. and Loria, R.M., Chapter 10, this volume.
14. Adi, F.C., Chapter 11, this volume.
15. Jenson, B. and Rosenberg, H., Chapter 9, this volume.
16. Ryder, E. and Ryder, S., Chapter 12, this volume.
17. Hayward, G.S., Ambinder, R., Cinfo, D., Hayward, D. and La Femina, R.L. J. Invest. Dermatol. 83:295-415, 1984.
18. Umene, K., Watson, R.J. and Enquist, L.W. Gene 30:33-39, 1984.
19. Rixon, F.I., Campbell, M.E. and Clements, J.B. J. Virol. 52:715-718, 1984.
20. Becker, Y. Med. Hypoth. 18:187-192, 1985.

Index